WORKSHOPS IN COMPUTING
Series edited by C. J. van Rijsbergen

Also in this series

ALPUK91, Proceedings of the 3rd UK Annual Conference on Logic Programming, Edinburgh, 10–12 April 1991
Geraint A. Wiggins, Chris Mellish and Tim Duncan (Eds.)

Specifications of Database Systems,
1st International Workshop on Specifications of Database Systems, Glasgow, 3–5 July 1991
David J. Harper and Moira C. Norrie (Eds.)

7th UK Computer and Telecommunications Performance Engineering Workshop,
Edinburgh, 22–23 July 1991
J. Hillston, P.J.B. King and R.J. Pooley (Eds.)

Logic Program Synthesis and Transformation,
Proceedings of LOPSTR 91, International Workshop on Logic Program Synthesis and Transformation, University of Manchester, 4–5 July 1991
T.P. Clement and K.-K. Lau (Eds.)

Declarative Programming, Sasbachwalden 1991
PHOENIX Seminar and Workshop on Declarative Programming, Sasbachwalden, Black Forest, Germany, 18–22 November 1991
John Darlington and Roland Dietrich (Eds.)

Building Interactive Systems: Architectures and Tools
Philip Gray and Roger Took (Eds.)

Functional Programming, Glasgow 1991,
Proceedings of the 1991 Glasgow Workshop on Functional Programming, Portree, Isle of Skye, 12–14 August 1991
Rogardt Heldal, Carsten Kehler Holst and Philip Wadler (Eds.)

Object Orientation in Z
Susan Stepney, Rosalind Barden and David Cooper (Eds.)

Code Generation – Concepts, Tools, Techniques
Proceedings of the International Workshop on Code Generation, Dagstuhl, Germany, 20–24 May 1991
Robert Giegerich and Susan L. Graham (Eds.)

Z User Workshop, York 1991, Proceedings of the Sixth Annual Z User Meeting, York, 16–17 December 1991
J.E. Nicholls (Ed.)

Formal Aspects of Measurement
Proceedings of the BCS-FACS Workshop on Formal Aspects of Measurement, South Bank University, London, 5 May 1991
Tim Denvir, Ros Herman and R.W. Whitty (Eds.)

5th Refinement Workshop, Proceedings of the 5th Refinement Workshop, organised by BCS-FACS, London, 8–10 January 1992
Cliff B. Jones, Roger C. Shaw and Tim Denvir (Eds.)

Algebraic Methodology and Software Technology (AMAST'91)
Proceedings of the Second International Conference on Algebraic Methodology and Software Technology, Iowa City, USA, 22–25 May 1991
M. Nivat, C. Rattray, T. Rus and G. Scollo (Eds.)

Humphrey Sorensen (Ed.)

AI and Cognitive Science '91

University College, Cork
19–20 September 1991

Springer-Verlag Berlin Heidelberg GmbH

Humphrey Sorensen, BE, MSc, MS
Department of Computer Science
University College, Cork
Ireland

ISBN 978-3-540-19785-0

British Library Cataloguing in Publication Data
A catalogue record for this book is available from the British Library

Library of Congress Cataloging-in-Publication Data
AI and cognitive science '91 : University College, Cork, 19-20 September 1991 / Humphrey Sorensen, ed.
p. cm. – (Workshops in computing)
"Papers presented at the 4th Irish Conference on Artificial Intelligence and Cognitive Science" – Pref.
"Published in collaboration with the British Computer Society."
Includes bibliographical references and index.
ISBN 978-3-540-19785-0 ISBN 978-1-4471-3562-3 (eBook)
DOI 10.1007/978-1-4471-3562-3
1. Artificial intelligence–Congresses. 2. Cognitive science – Congresses. I. Sorensen, Humphrey, *1955–*. II. British Computer Society. III. Series.
Q334.A443 1992
006.3-dc20 92-18611
CIP

Typesetting: Camera ready by contributors

34/3830-543210 Printed on acid-free paper

Preface

This book contains the edited versions of papers presented at the Fourth Irish Conference on Artificial Intelligence and Cognitive Science (AICS'91), which was held at University College, Cork, Ireland on 19–20 September 1991. The main aims of this annual conference series are to promote AI research in Ireland, to provide a forum for the exchange of ideas amongst different disciplines concerned with the study of cognition, and to provide an opportunity for industry to see what research is being carried out in Ireland and how it might benefit from the results of this research. While most of the participants at the conference came from universities and companies within Ireland, a positive feature was the interest shown from outside the country, resulting in participants from Britain, USA and Italy. The keynote speaker was Professor James A. Bowen, North Carolina State University, who spoke on future trends in knowledge representation. The topics covered in the presented papers included fundamental approaches to AI, natural language, knowledge representation, information retrieval, deduction, epistemics and vision. The sponsors of the conference were Digital Equipment Co. (Galway) and Eolas, the Irish Science and Technology Board.

March 1992 Humphrey Sorensen

Contents

Section 1:

Foundations and Methodologies

TOWARD A NEW FOUNDATION FOR COGNITIVE SCIENCE

Author: Sean O Nuallain
Chair, Computational Linguistics Programme Board,
Dublin City University, Dublin 9, Ireland

E-mail: ONUALLAINS@DCU.IE

Phone: 353-1-7045454

Abstract

This article addresses itself to the conceptual crisis in present-day Cognitive Science. It argues that, in particular, cognitive psychology and AI suffer from a destructive Cartesian legacy. Moreover, the modern schools of philosophy which have found fault with Descartes may provide a more secure foundation on which to build Cognitive Science. It is certainly over-sanguine to hope that the new techniques afforded by connectionism will on their own found a less problematic discipline. Merleau-Ponty's work on perception is examined in a philosophical, then a computational context. A line of argument is then developed wherein Merleau-Ponty's and Piaget's perspectives are synthesised. From this, a new perspective on the nature of subject/object relations in human cognition is developed, and related to current Cognitive Science. Finally, future directions for research are proposed.

1. Introduction: The Scope of Cognitive Science

It is fair to say that there exists a great deal of confusion about the precise nature of Cognitive Science (henceforth CS). Different sections of the academic community hold widely varying conceptions of its nature. For psychologists, it consists of that aspect of cognitive psychology which attempts to give a computationally-tractable account of mind. For computer scientists, it is at present vaguely identified with neural networks. For cognitive scientists themselves, it is a meta-discipline, including inter alia both cognitive psychology and AI.

For Gardner (1985) the scope of CS is quite clear: "(CS is)... a contemporary, empirically-based effort to answer long-standing epistemological problems" (p.6). Indeed: "Today, CS offers the key to whether they can be answered" (ibid).

Pylyshyn (1984) is even more explicit:

"Many feel, as I do, that there may well exist a natural domain corresponding roughly to what has been called 'cognition' which may admit of such a uniform set of principles" (Preface, P. xi).

For Pylyshyn, then, cognitive science is on a par with disciplines containing such a set (e.g. "chemistry, biology, economics or geology" (ibid)) in having a precisely-specified subject-area: "... the domain of cognitive science may be knowing things... informavores" (ibid).

Stillings et al (1987) are even more forthright: "Cognitive Science is the science of mind" (p.1). Like both Gardner and Pylyshyn, they stress the interdisciplinary nature of the enterprise of CS, and focus on the disciplines of "psychology, linguistics, computer science, philosophy and neuroscience" (ibid) in their account of the area. We shall discuss this issue of the disciplines involved presently. Johnson-Laird (1988), whose earlier incarnation was as a cognitive psychologist unsurprisingly gives a psychologist's definition: "Cognitive Science, sometimes explicitly and sometimes implicitly, tries to elucidate the workings of the mind by treating them as computations (p.9). The disciplines of "psychology ... artifical intelligence, linguistics, anthropology, neurophysiology, philosophy" (ibid, p.7) are seen as the most relevant.

Let us pause for a moment to take stock. There seems to exist a consensus on at least these basic issues:

i. Cognitive Science is the science which deals with knowing
ii. As such, it must accommodate findings from a variety of disciplines

This established, we now need to look at the modus operandi of CS. That accomplished, we need to examine what the relevant disciplines have actually told us about cognition.

With all this review finished, some themes will have emerged which run through all the disciplines involved. It will be established that there is indeed a need for a CS discipline (or meta-discipline) but that certain of the current assumptions of the area are greatly flawed. These assumptions will be made explicit, and their flaws pointed out. A new foundation for the area will be proposed. Finally, some fruitful directions for future research and their antecedents in current research will be indicated.

2. Methodological Considerations

Gardner (1985) supplies the most lucid characterisation of the methodology of CS. He predicates five features of CS (p.38 et seq).

(i) The acceptance of a symbolic level, of a level of representation. This point is amplified considerably by Pylyshyn (1984):

I will suggest that one of the main things cognizers have in common is they act on the basis of representations" (p.xii). It should be pointed out that this is the key-note of CS: if the representationalist ethos is accepted, then the domain becomes computationally tractable. If it can be established that humans work from symbolic representations, then the task of CS becomes that of implementing these representations as programs, and CS seems a priori quite a feasible enterprise. Without wishing to spoil the story, I should point out that representationalism has been attacked on philosophical (MacQuarrie, 1972), perceptual (Gibson, 1979), and methodological (Winograd, 1990) grounds.

ii) The use of computers. As Gardner (p.43) points out, CS could have emerged at any point since the 1940s, but acquired its initial credibility and impetus from the later success of computing.

iii) The playing down of the influence on cognition of affect, context, culture and history. This de-emphasisis can also be contested on various grounds.

iv) The interdisciplinary ethos, as we have already examined above. Gardner (p.44) describes CS as being bounded on its extremes by neuroscience and anthropology. Its core comprises cognitive psychology, AI and large amounts both of linguistics and philosophy.

v) Finally, Gardner rightly points out that CS is rooted in classical philosophical problems and dichotomies e.g. empiricism versus rationalism. Consequently, its relation to philosophy is Janus-faced. It must continually both look back at the past to avoid falling into any of the classical fallacies while it simultaneously remains seized of modern developments in semantics and linguistic philosophy, inter allia.

It is argued below, apropos these points, that the first and third are problematic. They are related to the second in that the use of computers for CS requires representationalism, but is incompatible with the incorporation of affect, context, culture and history. With respect to the fourth point, the interdisciplinary aspirations of CS are to be applauded, but great difficulties arise with formulating a language of explanation comprehensible to all parties involved. However, an attempt must be made to surmount these difficulties in particular.

As we proceed with the review of findings from the relevant disciplines, we shall notice several themes with variations in different settings begin to emerge. These themes are precisely those issues within philosophy on which CS must take a stance in order to fulfil its destiny as the Science of the Mind. It will be argued, in this context, that its current set of stances is ill-selected.

3. Relevant findings from the constituent disciplines of Cognitive Science

3.1 Introduction

Gardner's conception (noted above) of the inter-disciplinary structure of CS will be accepted for the purposes of the forthcoming account. Even without the common platform supplied by CS, the disciplines in question shade into each other quite naturally and it is sometimes necessary to fudge the boundaries. For example, the psychologist's mission to explain the processes by which mind performs a specific task finds an echo in the computer scientist's desire to design and implement an algorithm which can do the same task. At a more fundamental level, Wittgenstein's early "logical atomism" (Ayer, 1982, p.111; Wittgenstein, 1922) program is precisely replicated in the latter-day AI "compositional semantics" enterprise (Charniak et al, p.225). Moreover, Wittgenstein's (1967) thoroughgoing critique of logical atomism may be said to be valid both for compositional semantics and even Fodor's notion of a unitary "language of thought" (Wilks, 1990).

To complicate the situation, some AI researchers, flush with the success of their systems in micro-worlds, feel disposed to hold forth on topics as diverse as selfhood, consciousness, the nature of the soul, and freedom of the will (Minsky, 1986 is a fine example) while holding fast to their original, restricted set of computational considerations. Winograd (1990) argues that this is a dangerous trend. Nor is this disease confined to computer scientists. Johnson-Laird (1988) has an equally wide range of interests and an equal lack of appreciation for the complexity of the issues he raises. Indeed, in true Messianic style he offers eternal life through simulation of personality (p.392). The broadcaster and neurophysiologist Blakemore (1988), in his popular TV series on the mind, within the space of two pages (pp.270-272) first of all emphasises the autonomy of moral issues, and then claims that they also fall within the domain of neurophysiology. We shall have little more to say on such misguided notions as those of Minsky, Johnson-Laird and Blakemore. They relate to philosophy only in that the errors they make are clasical philosophical errors.

In fact, each of the areas we are about to discuss is in itself enormous, and can only be done justice to by an appropriate University course. Obviously, we shall consider only major trends within each. Finally, each is to be dealt with in its own terms, before being put in a more encompassing CS context.

3.2 Cognitive Psychology

Bolton (1979) claimed that psychology "is capable of a crisis in its basic concepts" (p.174). It is indeed possible that the success of cognitive science in the 1980s prevented this crisis from being noticed. For Bolton, the crisis was to be a healthy one, a re-evaluation of the basic concepts of his discipline. Let us examine the strands in Bolton's argument.

Most of all, Bolton emphasises the dnagers of "psychologism" i.e. the trivialisation of autonomous formal systems like those of mathematics by their reduction to matters of subjective judgement. Yet this subjectivism is the inevitable consequence of current thinking in the relationship between the subject and his world.

If, as is claimed commonly in cognitive science, we build representations only on the grounds of their heuristic usefulness, and these representations need not necessarily faithfully reflect the corresponding reality (Stillings et al, p.306, go this far), then no objectivity can exist.

Bolton resolves this question by stating that psychology must become phenomenologically based. I have outlined this argument elsewhere (Nolan, 1991). Essentially, one lesson must be learned from Bolton's analysis: as they stand, both CS and psychology have difficulty accommodating the objective nature of formal systems into their explanatory vocabulary.

Piaget (1926, 1970) had much to say on the construction of this external world. His remarkable lifetime project was an explication of the processes whereby it is built up. However, the "epistemic subject" (Furth, 1981) Piaget regarded as co-created with that subject's world. We cannot do anything resembling justice to Piaget's system here: what is relevant are the questions of genesis of the subject and her world raised, rather than Piaget's specific claims. In this, Piagetian vein, Morgan (1979) describes our concept of space in terms of physical action.

So far, then, we have examined psychology's notion of the external world. We cannot proceed on this path without commenting on consciousness which relates us to the world. Thines (1977) describes this relation in the vocabulary of phenomenology. Wetherick (1979) shows that one can, with some consistency, argue that the individual consciousness is either non-existent, all-encompassing, or best described only in terms of its objects, before deciding on the last alternative (p.108).

Up to this point, we have considered epistemological issues in psychology at rarefied, abstract levels. Neisser (1976) can help us return to earth. On the one hand, he insists that psychology should have "ecological validity" i.e. it should inform our understanding of how real people act in real situations. On the other hand, he frequently refers to his then Cornell colleague J.J. Gibson and the latter's concern with "An ecological approach to visual perception" (Gibson, 1979).

Gibson's Weltanschauung is as far away from that of CS as can be without leaving psychology altogether. For Gibson, the contemporary notions of representation were absurd: his environment was already full of information to be picked up, unmediated by representation. Such CS researchers as Fodor et al (1981), and Pylyshyn (1984) objected vehemently, pointing out that the "affordances" which Gibson proposed were unconstrained. In reply, the Gibsonians (e.g. Turvey, 1977) insisted that the contrasting view put the same onus on inference as their own put on affordance.

This controversy - it is beyond a mere debate - will run and run. I wish now to propose a synthesis, the motivation and rationale for which will become clearer as we proceed. Essentially, both sides are partially correct. As the Gibsonians insist, we are adapted in specific ways to the environment, and show a sensitivity to higher-order invariants therein (e.g. for movement) which cannot be described in the ordinary representationalist schema. However, even these invariants must be processed in some way, which re-introduces, at least partially, the "Establishment" metaphor.

If pressed to its logical conclusion, Gibson's theory states that this page affords reading about Cognitive Science, without paying due attention to whether the subject knows the area, or even how to read. Likewise, Fodor et al's (1981) viewpoint, as they are themselves aware, posits an enormous set of inferences, for any action or more complex that the reflex-arc. Again, we see that psychology has difficulty in relating the subject to the external world in any coherent fashion. Representationalism can explain the difference between an infant's and an expert's knowledge of a book, but not the perception of a world of enduring objects, for the ecological approach, the situation is precisely the opposite.

What is clear is that this either/or dichotomy is unacceptable in psychology and some via media must be found which can preserve both the objectivity of the external world and the possible preparation of a subject to perceive some entities rather than others.

3.3 Artificial Intelligence

3.3.1 *Introduction*

It was mentioned above that the contemporary success of the CS approach temporarily defused a time-bomb in cognitive psychology. In particular, due to the commercial breakthrough of expert systems, the late 70s and early 80s were golden years for AI, and consequently for CS. However, the time-bomb is again primed, and occasionally looks like blowing up AI as well as cognitive psychology.

That there is a conceptual crisis within AI has become clear even to the editors of Byte magazine, who recently saw fit to publish a set of articles exploring the foundations of AI (Ryan, 1991). However, this is by no means the first time that AI has reached such a juncture (Lighthill, 1972), or that a specific topic within AI has received a premature obituary (Pierce, 1966; Minsky et al, 1969). It is worthwhile, with this in mind, briefly to review the history of the area.

The early history of AI is a series of spectacular successes on such areas as theorem-proving (Newell et al, 1972), pattern-recognition using connectionist methodology (Rosenblatt, 1962) and machine translation. What these disparate activities had in common were:

(a) a treatment of the subject-area in terms of the regular interaction of formal symbols.

(b) a recalcitrance to take on problems in the real world. A soon as reality intruded, the fragility of these systems became evident. Yet some applications can be implemented using this fragile methodology: for example, an expert system for spetrocsopic analysis requires only this syntactic level (Fergenbaum et al, 1983).

Following the short, sharp shocks administered by reports such as those by Pierce and Lighthill (opera cit), AI researchers began to represent knowledge at the level of structured objects, rather than syntax. Systems such as KL-One and KRL (Brachman and Levesque, 1985) made their appearance. Simultaneously, Schank et al (1977), adapting the psychologist Frederic Bartlett's (1932) notion of "schema" produced "semantic-level" representations for stories. In the expert systems field, Duda et al (1984), inter alia, produced structured object formalisms for particular domains like geology. It is worthwhile mentioning that all these second generation knowledge representation systems have in common a definite context-specificity. Neither Duda et al's nor Schank's higher-level formalisms can be transferred whole from domain to domain, as Newell et al earlier thought might be possible for general problem-solving principles.

We shall look presently at precisely what now seems possible in AI. For the moment, it is worthwhile to note what Gardner terms the "Computational Paradox" the fact that the application of computer models to human thought has told us as much by its failures as by its successes. The recent re-birth of connectionism has given many new hope.

With reference to this, Anderson (1988) points out "Early on, one had a choice of (at least) two paths to processing: the digital computing path and the neural network path. Digital electronics was the more promising path. Now that we see some tasks are not so easy the one way, it is time to explore the other one" (Introduction (p.v)). She summarises her argument thus:

"What has changed is our perspective: we have begun to see the limitations of the digital computer. Indeed, we have begun to run into them (ibid)". Therefore, we see that neural networks look inevitable, even from a Anderson's purely technical point of view. We shall shortly consider them from an epistemological perspective.

For the moment, I propose to deliver on the promise of a review of some key areas in AI.

3.3.2 *Natural Language Processing (NLP)*

Winograd's (1972) SHRDLU was the turning point for NLP. It demonstrated the following

i) Success was possible for NLP in a restricted context
ii) However, this context was a "micro-world" and it became obvious that it was a formidable task to develop context-general principles.

So the situation remains. The algorithms have become more efficient (Tomita, 1986); the applications include NL interfaces of increasing coverage (Hendrix, (1987) and interlingua-based machine translation systems of ever greater sophistication (Nirenburg et al, 1987). Yet even these successful systems remain tied to specific domains and there does not seem to be any principled ways of fitting a system to a domain; much is left to ad-hoc improvisation and trial-and-error.

Where language needs to be considered only on the syntactic level, the task of building a parser seems straightforward (e.g. for grammar-checking).

i) If the context is restricted sufficiently to allow conveyance of semantic relations by refined syntactic formula (e.g. data-base interface), the requisite system can be created. However, that system will include much that is domain-dependent.

ii) If the context is less restricted than this, the domain may be mapped by semantic primitives. Yet again, the semantic primitives selected will contain several which are domain-dependent (Schank et al, 1977).

iii) When the text understanding requires speech-act sensitivity i.e. a knowledge of the speaker's intent, a less-than-graceful degradation of performance occurs (Winograd, 1990). It is proving very difficult to handle context change.

In summary, then, NLP system work well if the domain can be mapped syntactically, and second-generation systems work with semantic representations within specific domains. As yet, no principled methodology exists for moving between domains, nor for handling speech-acts.

3.3.3 *Expert Systems*

Expert Systems remain AI's major success story and this require little introduction. Again, it is appropriate to distinguish between:

i) Systems which map the domain at a syntactic level. This is quite sufficient for rule-book applications (Fergenbaum et al, 1983)

ii) Systems which attempt representation by structured object representations e.g. the graph theory application in Internist and Prospector's inference netowrks (Duda et al, 1984).

iii) The intractable problem of context-independent systems which map all possible domains at a fundamental level

3.3.4 *Vision*

The situation for vision is neatly summarised by Tanimoto (1990):

"A computer system has not yet been created which can look at an image and describe the scene depicted" (p.410). That goes even for static images. Movement has proven even more difficult to handle (Thompson et al, 1990). Again, we can distinguish between

i) A description, like Marr's "primal sketch" notion of the "syntax" of the scene, which has proven to be a tractable problem.

ii) The much more difficult task of scene description (Tanimoto, op cit).

iii) The currently intractable problem of movement.

3.3.5 *The Frame Problem*

Pylyshyn (1987) identifies the frame problem with the problem of relevance:

"This, the problem of relevance, is what many believe lies at the heart of the frame problem" (Introduction, p.x). Hayes (1987) in the same volume, indicates its aetiology:

"The frame problem arises when the reasoner is thinking about a changing, dynamic world, one with actions and events in it" (p.124). Essentially, it relates to the processing of change by a system based on updating of representations for handling change. No principled methods exist of determining what is relevant or otherwise in any change which occurs.

Still (1979) considers that Gibson's system provides a solution to the frame problem at the perceptual level. For Gibson, the actual stimulus is the very continuous perspective transformation that AI systems find so difficult. (As a point of interest, Berkeley also foresaw the frame problem, and produced a theological solution of it whose day may yet come!).

In essence, the frame problem is inescapable for any representationalist system, and it is as well to stress this point now. In fact, if we broaden the concept of "change" to include on the one hand change of domain for expert systems and NLP, and on the other movement for vision systems and change of speaker for speech-recognition systems, then the frame problem acquires a new, disturbing life. It then seems to reflect a much more fundamental difference between human cognition and AI than hitherto suspected: in fact, it threatens the representationalist paradigm as a good model for the one, and as a technical apparatus for the other.

Some within the connectionist community have argued that theirs is a new paradigm for AI:

"The upshot of this is that connectionism provides a kind of Copernican revolution in cognitive explanation. Instead of having the running system revolve around an antecedent analysis of the task, we may make the analysis of the task revolve around an up and running system" (Clark et al, 1990, p.12). The domain will map itself onto the system by a "continued process of analysis" (ibid). Gardner (1985) comments on the closeness to Gibson's system of PDP (pp.321-2) but refrains from taking sides on the issue of whether connectionism can provide a new framework for AI.

There exists at least a possibility that PDP may not fall foul of the frame problem in precisely the same way as classical AI. Let us now briefly survey the area covered by PDP.

3.3.6 *Connectionist systems (PDP)*

The initial auguries are unpropitious: "Although the current generaiton of networks appears to be very good at performing/learning to do what are essentially static/spatial tasks, there has been relatively little progress on networks that can cope adequately with tasks which have an important temporal/sequential component to them" (Clark et al, 1990, p.11). Perhaps PDP can in the future evade the frame problem; at present, such is not the case. Moreover, the claim for neurological veracity has been abandoned; PDP must stand or fall on its own merits as a cognitive model (Wilks, 1990). Perhaps the only truly novel feature of PDP which sidesteps an old trap-door is its putative subsymoblic nature but even this is hotly contested by Fodor (Wilks, ibid).

These PDP systems which attempt ecological validity are vulnerable to the classic critique of behaviourism (Perschl, 1990). In all, the successes of PDP in pattern-recognition and cognitive veracity must be set against its current failures.

3.3.7 *Summary*

We are left with the following picture of AI:

i. Many ingenious methods have been developed to treat various domains at a syntactic level for the modalities of speech and vision, and for reasoning

ii. Where a real-world application for a specific domain is required, the mapping always involves considerations relevant only to that domain

iii. No principled way has yet been found of handling domain change.

3.4 Linguistics

Section 3.3.2 is concerned with one aspect of Linguistics i.e. Applied Computational Linguisitcs (ACL) a synonym for NLP. It is worthwhile examining the provenance of this term.

Linguistics may be considered as the attempt to give a formal characterisation of a Language, or in a chomskyan context, of a Grammar. Computational Linguistics is the study of how to transform this characterisation into a computationally-tractable description. ACL is the use of this description as an instrument toward a specific end (Thompson, 1983).

Where linguistics relates to the concerns of this paptor, it has already been dealt with in 3.3.2 above. However, it is worthwhile to flag two salient facts

a) The Chomskyan transformational model no longer holds sway and has not done so for several years (Smith et al, 1979). Yet it continues to influence AI due to its computational tractability.

b) The concerns of the theoretical and AC linguist have diverged sufficiently to give rise to the autonomous computing field of language- engineering.

In short, we have woken from the dream of a comprehensive computer-tractable formal description of linguistic competence and performance.

3.5 Neuroscience

Along with anthropology, this area is one of the boundaries of CS. It is worthwhile simply noting the following points:

i. The "neuromorphist" claim has, as we've seen, been abandoned by connectionists (Wilks, 1990)

ii. Only rarely can an algorithmic description for the action of specific neural groups be informatively given (Churchland, 1990 gives one such).

iii. The equation mind=brain + programs, beloved by researchers like Johnson- Laird (1988), p.7) and Pylyshyn (p.258) is by no means borne out by neurophysiological research. Eccles (1987), and Cray (1987) argue against monism of any description, and Eccles at least opts for a wholesale dualism.

iv. Such notions as consciousness and selfhood remain untouched by neurophysiological explanation.

3.6 Anthropology

Again, this other boundary for CS has little more than caveats to offer for researchers eager to give a computatioally tractable description of mind. Among the central findings of this attempt to study man, as it were, in situ is:

"Neither language nor culture gives rise to consciousness" (Gardner, 1985, p.235).

The brief accorded to ethnoscience should give pause to those same researchers:

"Ethnoscience is the organised study of the thought systems of people in other cultures and our own" (ibid, p.244).

Including, one hopes, the thought systems of CS researchers.

3.7 Philosophy

3.7.1 *Introduction*

Ayer (1982) makes some illuminating comments on the role of philosophy in contemporary scientific and other discourse. He remarks, first of all, that it lacks the capital to set up shop on its own: the days of

Leibnizian system-builders are long past. Science often encroaches on philosophy, and one need only refer to Aristotle's "physics" to realise how far these incursions have gone. Secondly, he notes that some philosophical positions, like absolute Idealism, cannot be refuted, but must be abandoned, rather like one abandons a sinking ship.

It must be added, and this time contra Ayer, that there exist instances where psychological analysis must take priority over that by philosophy. For example Ayer remarks with reference to the work of Merleau-Panty on the nature of the subject which we will shortly consider:

"Psychologically, this may well be true. Philosophically, it may be held to beg the question" (Ayer, 1982, p.221). Frankly, in the particular case of the nature of the subject the psychological truth will suffice for our purposes.

Philosophy can help greatly, however, in articulating precisely the assumptions on which CS is based. For example, Bolton (1979) castigates contemporary cognitive psychology for its basis on British empiricism, and offers phenomenology as a more secure basis; Winograd (1990), faced with the same choice in foundations for AI, chooses similarly.

Now, it is worthwhile briefly to review our findings so far. We learned form linguistics, neuroscience and anthropology that no single alchemist's stone exists in any of these disciplines which can convert are tentative graspings into golden certainties. We found that the frame problem is common to cognitive psychology and AI. However, the problem which the former has in explaining logico-mathematical certainty is explicable in the programming ethos of the latter.

It is useful to start our analysis of the philosophical foundations of CS by examining the basis proposed by Stillings et al (1987). They suggest the following three canons for CS (pp.305-6).

(i) Representationalism
(ii) Methodological solipsism
(iii) Dualism re-cast as functionalism

The re-casting in iii does not yet allow this formulation to escape the single most consequential argument in contemporary epistemology, an objection to Cartesianism which is neatly put by MacQuarrie (1972):

"If one begins with knower and known as separate entities, then perhaps it just is impossible to put these two together or to get out convincingly from the solipsist trap" (p.127).

He continues by stating the alternative scenario:

"The existentialist, as we have stressed, begins not with two things but with the unity of being-in-the-world" (ibid).

Following Steiner (1978), we shall term this existentialist perspective "mundanity" and contrast it with the "Cartesianism" on which CS is based. For the remainder of this section on philosophy, I propose to:

i. Highlight the contrast between the two perspectives in question by showing how they relate to specific issues

ii. Outline Maurice Merleau-Ponty's (1962) work on the "Phenomenology of Perception" which provides an encompassing non-Cartesian theoretical framework for a possible new direction for CS

iii. Indicate the relevance of Merleau-Ponty's work to certain continuing controversies in CS.

iv. Show that Passmore's (1967, pp.502-3) questioning of Merleau-Ponty's system is motivated similarly to Fodor's (1980) now classic notion of the innate "Language of Thought". I shall indicate how the objections of the former can be met and the innatism of the latter can be countered. At that point, I shall introduce my own views on the subject.

3.7.2 *Mundanity and Cartesianism*

It goes without saying that these are polar opposites on the central epistemological questions. For Kant, who intuited the limitations of

Cartesianism, the scandal of philosophy wa that it could not prove the existence of the external world: for Heidegger, the greater scandal was that it should feel compelled to do so.

With that, let us consider the first pillar of Cartesianism (which also supports the edifice of present-day CS) i.e. representationalism. Essentially, this is central for Cartesianism, and extra-mural for mundanity. The latter point of view urges that man is first and foremost Dasein, a being-in-the-world (MacQuarrie, op cit, p.125), and that the notion of representationalism is a damaging one, distancing us from our world. Obviously, this argument has environmentalist overtones for modern ears, spelled out in (Nolan, 1991).

Secondly, if as Merleau-Ponty (1962) argues, we are ab initio "situated in an intersubjective world" (p.355), solipsism even adopted as a methodological strategy is a mistake:

"The second interesting implication of Descartes's is that it is possible to study the mind without paying any attention to the reality it purports to represent and talk about" (Stillings et al, p.306). Frankly, that statement rings out, and the failure of AI systems to function in a dynamic environment adds a harmony line.

Thirdly, when recast as functionalism, Cartesian dualism becomes a parody of itself. The frantic attempt by contemporary CS researches to impose monist view while preserving the other trends of Cartesianism has an ultimately comic aspect (e.g. Johnson-Laird, 1988, p.392). Instead of a homunculus in the pineal gland, one is left with the absurdity of a Tuning Machine performing much the same functions on the requisite representations from the various modalities.

This point is worth dwelling on, because it shall presently be represented in a different context. Let us briefly note Heidegger's views on the subject of Dasein, which to start with, we shall relate to the first two pillars of Cartesianism before again commenting on dualism.

"It is not the case that man 'is' and then has, by way of an extra, a relationship-of-Being toward-the-world" (Heidegger, 1929). There exists no homonculus or separate Turning Machine frome the world at the most fundamental level. Steiner (1978) elucidates this point:

"To be human is to be immersed, implemented, rooted in the earth, in the quotidian manner and matter-of-factness of the world" (p.81).

Obviously, Heidegger's is a very strong anti-Cartesian line on representationalism and solipsism. Nor is he committed to dualism:

"Heidegger's 'mundanity', to use this eroded word in its strongest etymological sense, would overthrow the whole metaphysical mind-body tandem and the dissociation between essential being and being here-and-now" (Steiner, 1978, pp.82-3).

Merleau-Ponty's work is the major recent statement of the existentialist approach to perception. As such, it is worth examining now.

3.7.3 *The Phenomenology of Perception*

An appropriate point to introduce Merleau-Ponty is his argument that I (the epistemic subject which he terms the "body subject") could not prevail over my subjectivity "if I had not, underlying my judgements, the primordial certainty of being in contact with being itself, if before any voluntary adoption of a position, I were not already situated in an intersubjective world" (Merleau-Ponty, 1962, p.355). As Ayer (1982) acidly notes "The only adults for whom the perception of other people and the intersubjective world are problematical are either deranged or engaging in philosophy" (p.221). This intersubjective world resembles Gibson's rather than Fodor's:

"Our perceptual field, Merleau-Ponty argues, consists not of sensations but of things with spaces between them" (Passmore, 1966, p.500).

Not only that: it comprises also other people, with intentions similar in structure to one's own. Here, then, is a basis for a meaningful cognitive psychology (Thines, 1977).

Merleau-Ponty's framework resembles Heidegger's in its refusal to play down the role of the body in cognition. Indeed, his notion of the epistemic subject he terms the body-subject. Passmore (1966) expounds at length on the body and the body-subject. In certain contexts the body is perceived as an object, in other contexts it is the perceiving subject, "The body-subject always perceives more than 'is there', more than can be analysed in terms of light-waves and retina; it sees houses, not facades; it perceives what lies behind as well as what lies in front" (p.501).

With this Gibsonian statement, we shall now return to the more travelled roads in cognitive science.

3.7.4 *Merleau-Ponty, Gibson and AI*

There are no explicit references either to Heidegger nor to Merleau-Ponty in Gibson's work, which makes the confluence of themes all the more remarkable. Gibson (1979) is very much a proponent of the "mundane" notions of mind/body relations; moreover, the specifics of his concerns and those of Merleau-Ponty often coincide:

"When it comes to the visual field Merleau-Ponty maintains, I think rightly, that it is given as three-dimensional" (Ayer, 1982, p.222). This perspective on perspective is shared by Gibson and is one of the many virtues even Pylyshyn (1984) sees in the latter's work. Moreover, Merleau-Ponty would not have found anything odd in Gibson's notion of affordances (pace, Fodor and Pylyshyn, 1981): "The body-subject encounters a world which already has 'meanings' incorporated in it" (Passmore, 1966, p.502).

The fact that we have no coherent neurological account for processing of these meanings should not obscure their reality, nor should the difficulty of giving them a computationally tractable description. Nor, finally, should the Cartesian mindset in which we are all to some extent fixed prevent acceptance of these entities. Let us bracket for the moment the question of how they could be processed. Despite its current computational intractability, this account of perception is closer to the facts than the Cartesian account. If as much research time in AI had been occupied in describing these "meanings" or Gibson's "invariances" as has been allocated to the Cartesian program, perhaps the area would have progressed more quickly.

There exist, also, higher-level entities for NLP, which are handled by speech-act theory (Austin, 1962). Winograd (1990, Winograd et al, 1986) comments that the consequences for NLP are that it must either address itself to the social context of language or else remain restricted to its current capabilities. Again, we must take account of intersubjectively-validated "meanings".

Philosophically, then, Merleau-Ponty's position is more tenable than the Cartesian one. It is strikingly analogous to Gibson's and suggests an alternative CS research paradigm. Again, like Gibson's system, it is not without its critics.

Let us now examine one such critique, before outlining an alternative, extended such position.

3.7.5 *A critique of Merleau-Ponty*

Like Gibson, Merleau-Ponty failed fully to explain "how cognitive truths could be derived from perceptual experience" (Passmore, 1966,, p.503). His fundamental account of perception we shall take as valid. Piaget (1926, 1970, 1972) can solve the above problem of derivation; in essence, it happens through action.

In fact, Merleau-Ponty's and Piaget's systems fill out the gaps in each other remarkably well. On the ultimate epistemological level, Piaget needs the kind of subject-object ambiguity of the body which Merleau-Ponty provides; conversely, Merleau-Ponty needs the kind of experimental psychological evidence, as well as the relation of cognition and perception through action, which Piaget provides.

It has not perhaps been emphasised enough quite how radical Piaget's epistemology is. In the rush to replicate (or not) his experimental findings, the baby found himself lying on the street alongside the bathwater. What are more relevant in Piaget are than the specifics of his experimental findings are:

The notion of the constructon of a subject and of a world through action
The idea that learning can also take place by differentiation of subject from object, self from world, as well as the more prosaic processes which can be modelled by adding facts to a database.

With this latter point in mind, we can now address ourselves to the second aspect of Passmore's critique, and with it Fodor's (1980) central argument: "How can we add to the stock of objective truths on the basis, merely, of new 'lived experiences'? How, again, is it possible to say anything at all about such experiences - to construct a phenomenology of perception - without making those sharp distinctions between ourselves and the world, which Merleau-Ponty does not allow? (Passmore, ibid, pp.502-3).

4. A new perspective on the epistemic subject

I wish at this point to introduce my own line of argument, which takes off from precisely this point. I wish to argue that these sharp distinctions are the exception, rather than the rule, in cognition, but occur nevertheless. In fact, on their rare occurrence in human cognition, the Cartesian representationalist standpoint is correct. Cartesianism actually is a limit-case for cognition: at moments of discovery or directed work, the Cartesian notion of the subject-object relation is (momentarily) correct. Let us look at one instance: Piaget's (1926) description of the infant's notion of the physical world. In essence, there is no differentiation as yet between the subject and object, and the infant projects its subjectivity onto the physical environment. With its initial coordination of actions, the child learns the requisite differentiation in the context of these actions.

Piaget continues his analysis with the cognitive development of the older child. Again, the crucial stages of development are those at which the subject becomes differentiated from the object: for example, the child may learn to distinguish the true nature of the experimental apparatus from the set of operations he can perform on it.

Merleau-Ponty's work provides a hospitable foundation for Piaget's. The body we noted, could be either subject or object, depending on the context. "Thus experience of one's body runs counter to the reflective procedure which detaches subject and object from each other" (Merleau-Ponty, 1962, pp.198-9).

Therefore, the body is ambiguous in its metaphysical status. Piaget supplies a schema in which the body-subject can learn. We can now address ourselves to Fodor's (1980) radical innatism.

Essentially, Fodor claims that for the learning of a new concept to occur, it is pre-requisite that the subject must already possess the concept in embryonic, pre-articulate form. A consequence of this view is the notion that there exists an innate "language of thought", a primeval inherited network of pre-articulate conceptual knowledge.

It is indeed true that the Cartesian framework cannot allow true conceptual learning, and with respect to that Fodor's argument is another nail in its coffin. If learning is to be restricted to the updating of a representation or even of a principle of representation, then Fodor's argument is valid. If, however, learning can change the relationship of subject and object at the most fundamental metaphysical level, then Fodor's argument no longer holds.

Fodor confuses the epistemological subject with the essential self, and this error, too, is a Cartesian one. The epistemological subject may be co-transformed with the objective world and a new level of adaptation established.

It is worthwhile parenthetically to note that other cultures refuse to make the Cartesian distructions that are inevitable for us:

"Japanese sometimes have feelings of superiority toward Westerners, who in their eyes cannot easily become one with nature. The Japanese never experience the splitting of the body and the sould that occurs in the consciousness of Western people ... nor (are they) burdened by the severe dichotomy of subject and object that is inherent in the Western philosophical tradition" (Doi, 1986, p.155).

We find the hugely complex notions of selfhood, subjecthood and the "soul" being compiled into one simple solution in the Cartesian tradition. This is obviously an over-simplification: I argue in the next section that it is both a misleading and a dangerous one.

To introduce another strand of this argument, it is worthwhile referencing two of Piaget's most disputed findings. Essentially, he argued that children are always, and adults never, egocentric. This is patently untrue (Donaldson, 1979); both adults and children are occasionally egocentric; what change, most of all, are the things about which we are egocentric.

It is mentioned above that non-egocentrism is always achieved in a certain context. A child may differentiate the physical self and world with respect to a certain context of coordination of action (e.g. feeding; feeling and withdrawing from, pain) and fail to achieve this differentiation in other contexts. Some contexts remain with us almost continually; the intersubjective context of mutual perception begins in early infancy (Donaldson, ibid) and continues with added mediation of language through to adulthood.

Yet, for other contexts, non-egocentrism is a remote achievement. If one is deeply involved, in a co-operative enterprise, particularly as leader, it can be very difficult to conceive of it in one's absence. On a more prosaic level, how often does one have to be reminded "It's your turn?"

This account seems paradoxical precisely because we are immersed in Cartesianism. For example, the distinction made above between "self" and the "subject" is one which cannot be made within Cartesianism either in its pristine (Descartes, 1649) or its modern (Eccles, 1987) forms. Moreover, the fact that one can be egocentric with respect to certain domains and not to others escaped even Piaget. Finally, the distinction of an epistemic subject which is attuned to various domains and not others. and an essential "self" is not one that comes easily to Cartesianism.

A particular attunement of the subject to a domain can be handled conceptually by Cartesianism and technically by AI system for expert diagnosis, perception or conceptual analysis. What distinguishes human from any artificial cognition developed to date is precisely the ability to switch from domain to domain. In the domains mapped by linguistic knowledge, the human success is shared by all language-users; where the domain is a specialised human activity like expert diagnosis, precise domain models lead to technical success for AI systems within those domains.

Consequently, Cartesianism and with it AI systems are accurate at the boundaries of cognition, where a task or a learning process requires the existence of the "sharp distinctions between ourselves and the world". Passmore (op cit) mentions. However, the majority of human cognitive acts do not conform to this schema:

i. At a fundamental level, we are attuned to a world of things and other people (Merleau-Ponty, 1962; Ayer, 1982, p.221; Gibson, 1979).

ii. We can change between domains and modalities in a manner which is not yet understood. The next section includes speculation on this matter of change using the connectionist notions of content-addressable memory.

iii. The relation between ourselves and the world shows a spectrum from compete immersion therein through to complete differentiation thereof. Moreover, we may be at radically different points on this spectrum with respect to many different domains at any particular moment. Again, an overall model is proposed in the next section.

iv. Where the differentiation is absolute, it can be implemented as a top- down system.

v. The dream of general-purpose AI principles is one from which we must awaken. Each domain will have considerations (encoded in e.g. particular semantic primitives or inference networks) peculiar to that domain.

vi. Generally-applicable principles (like edge-detection or syntactic analysis) cannot inform of the meaning of the particular domain, which require the knowledge encoded by internalised action (Piaet, 1926, 1958, 1970, 1972).

vii. We have thus preserved both the ecological orientation of the Gibsonian system and the implementation-orientation of Cartesianism for particular domains. Gibson's (1979) own later work was an attempt to find the "right conceptual level" and it can be said with fairness that he did not quite succeed: his system does not allow even for the possibility of the top-down AI-systems which have proven feasible.

viii. To understand why the subject chooses particular domains rather than others in which to develop is to enter the conative area. No coherent CS can ignore forever the question of the will.

ix. Finally, the monism/dualism debate should be abandoned. It is impossible to talk about matter without introducing the subject through the notion of "observer status" (Penrose, 1987) or the cosmogonic notion of the "anthropic principle" (Hawking, 1989). At the level of the physical description of the brain, Penrose (1987) points out that observer status must be introduced; the neurophysiological evidence for monism versus dualism is ambiguous in the extreme (Gray, 1987); in philosophy "the question what is implied by saying that one and the same event has both mental and physical characteristics still waits for a sufficient answer" (Ayer, 1982, p.190).

What is essential is that the notion of moral responsibility must never by called into question again by pseudo-explanations of moral choice like those of Minsky and Blakemomre (opera cit). One of the most bizarre scientific coincidences of our time is that psychology abolished the observer in the same decade that physics re-introduced it (Hawking, 1988). Szentagothai (1987), an avowed materialist, makes precisely the same strong argument, contra Minsky et al, concerning moral responsibility.

The undoubted empirical experience of one's and other's selves brings back the intersubjective domain into cognitive psychology, where it is badly needed. Thines (1977) points out that there are vast amounts of consensually-validated experiences which psychology does not at present touch.

The existence and mysterious nature of consciousness is rarely nowadays denied in CS (Pylyshyn, 1984; howver, Johnson-Laird, 1988 provides yet another pseudo-explanation). Nor should such notions as Hussel's "Transcendental consciousness" be, a priori, dismissed; if the body can be subject or object, then the focus of consciousness is in purely metaphysical terms labile.

5. The Demise of Cartesianism: A Review with Consequences for AI

Let us now review the argument of this essay. We found that both Cartesian philosophy and cognitive psychology have difficulty with finding the correct voaculabary to express subject-object relations. We found that AI systems have difficulty in moving between domains which are mapped at a deep level.

The difficulties we found with the Cartesian position were the following:

a) It prevented us from giving such higher-level perceptual phenomena as "invariances" adequate attention. This is because the Cartesian subject is isolated from his environment. Thus, the methodological suggestion from our analysis for perceptual studies is that we should pay much more attention to the attunement of the subject to his environment (Gibson, 1979).

b) On a metaphysical level, the Cartesian system allows us only absolute subject/object differentiation (even Kant's notion of categories is unused). This, however, is a limit case for cognition. It is possible that at any given moment only "focal" awareness can be considered in this way, while "Tacit" awareness" (Polanyi, 1958) may best be modelled by context-addressable associative memory networks which at that moment have under-determined content. It is indeed possible that the brain works in parallel both in symbolic and sub-symbolic modes (suggested by Johnson-Laird, 1988). The symbolic mode corresponds to focal awareness, the sub-symbolic to the under-determined systems noted above. When any of the sub-symbolic systems becomes determined by the addition of relevant content, it enters focal awareness. This suggestion seems to fit many of the facts of phenomenological analysis as well as the restricted successes of both top-down and connectionist AI systems.

c) On the ecological level, the Cartesian system is profoundly anti-environmentalist in spirit (Nolan, 1991) in that it exiles u from the world. Moreover, it posits a counter-intuitive and frankly absurd notion of the non-importance of the intersubjective world: as we have seen Ayer (1982) argue, this world is a psychological a priori. A final point here is that as it alienates us from the world, so it alienates us from ourselves. The attempt to find a Cartesian soul leading from behind in all our actions is nonsensical. Its worthwhile to take two quotes at random from last year's Booker Prize winner (Byatt, 1990) to see how far this notion has been exposed.

"Narcissism, the unstable self, the fractured ego, Maud thought, who am I?" (p.251).

"We know all sorts of other things too - about how there isn't a unitary ego - how we are made of interacting, conflicting systems of things" (p.267).

However, if we realise our mundanity, these insights are no longer troublesome, and the notion of unitary self independent of its mundane manifestations can re-emerge.

d) The notion of a unifying "Language of thought" or common language of representation on which the Cartesian homunculus can act was refuted by Wittgenstein (1967). Within each context, the relevant semantic primitives are linked to the external words or actions in idiosyncratic ways. It has become obvious that such is the case for language (Schank, 1977) and expert systems (Duda et al, 1984); perhaps such an analysis of space as Morgan's (1979) could perform a similar function for vision.

e) Finally, the subject-object differentiation which is the core of Cartesianism cannot handle these moments at which a subject realises that he/she is an object in a given context. That is the real significance of Heidegger's "thrownness" (Winograd et al, 1986). In fact, this capacity to see oneself in the larger picture, this "will to differentiation "may be the central motor of cognitive development. As we enter any refined context of action, it is with the knowledge of our possible objecthood. This is precisely the kind of relation to the world which escapes Cartesianism and can only be handled in a phenomenological context (Bolton, 1979). Piaget (1926) erred in saying that the child's liberation of his physical self from the "Uroboros" ensured self/world differentiation once and for all. This primal event is re-enacted on further physical and social levels all through life.

Having considered points a-e it should be evident that the possibilities afforded in this context for research work in content-addressable memory and in specifying appropriate primitives for particular contexts are vast: They are, I insist, precisely the issues to which AI should now be addressing itself. Let us examine the consequences for CS in general.

6. Conclusions

The main concerns here have been to preserve the notions of an intersubjective world for cognitive psychology, that striving for ecological validity of Gibson's (1979), Neusser's (1976) and Winograd's later (1990) work, our common-sense notions of the world and our part in it, and the technical possibility of AI. We found Merleau-Ponty to be an appropriate booster-rocket, and the final system affords a perspective in which all these concerns have a place.

With reference to the first two points, the rescue of the intersubjective world will allow psychology to ask meaningful, important questions about real human motivation and the nexus of social relations, inter alia. On a perceptual level, the re-introduction of the ecological approach should give rise to appropiate research on invariances and meanings whether expressed as affordances or not. On a social level, we may begin to speak with credibility about commitment in verbal acts as something that cannot be reduced to mere subjectivism.

The early objections of the "Dreyfusards" had to be ignored by AI which would have otherwise collapsed in a paroxysm of self-doubt. Given the current commercial success of the area, we can now perhaps take stock. The framework outlined above gives AI systems a role for simulation of human cognition, as a limit case thereof. Connectionism is not going to prove any kind of panacea, but is in itself a very interesting meta theory.

It may be in keeping with the rest of this paper (hopefully not!) that I leave common sense till last. Our common sense tells us that we are at times part of larger autonomous systems, at times on our own, that having trouble communicating with others is the exception, not the rule; that our awareness at any point is of varying domains to varying degrees; and that moral issues cannot be evaded. Frankly, even had the Cartesian program in Cognitive Science continued its early success, we would have found it hard to change these vital tenets of folk psychology.

Anderson, D (ed) (1988) Neural information, processing systems. New York: American Institute of Physics.

Austin, J (1962) How to do things with words. Cambridge, MA: Harvard University Press.

Ayer, A (1982) Philosophy in the Twentieth Century. New York: Random House.

Bartlett, F (1932) Remembering: A study in experimental and social psychology. Cambridge, England: CUP.

Blakemore, C (1988) The mind machine. London: BBC.

Bolton, N (1979) "Phenomenology and Psychology". In N Bolton (ed) Philosophical Problems in Psychology. New York: Methuen + Co.

Brachman, R and H Levesque (1985) Readings in Knowledge Representation. Los Altos, CA: Morgan Kaufman.

Byatt, A. (1990) Possession. London: Vintage.

Charniak, E. and D. McDermott (1984) Introduction to AI. Workingham: Addison-Wesley.

Churchland, P. (1990) "Representation and high-speed computation in neural networks". In D. Partridge and Y. Wilks (ed) The Foundations of Artificial Intelligence. Cambridge, England: CUP.

Clark, A. and R. Lutz (1990) Guest Editorial, AI + Society, 4 (1), pp.3-17.

Descartes, R (1649) The Passions of the Soul. In E Haldone and G Ross (trans) The Philosophical Works of Descartes, Cambridge: CUP (1967)

Doi, T. (1986) The Anatomy of Self. New York: Kodansha International.

Donaldson, M (1979) Children's Minds. Glasgow: Collins.

Duda, R. and C Rebok (1984) The use of inference networks in prospector. In Reitman (ed) AI applications for business. Norwood, NJ: Ablex.

Eccles, J (1987) Mind and Brain; One or Two". In C. Blakemore and S. Greenfield (eds) Mindwaves. Oxford: Basil Blackwell.

Feigenbaum, E. and P. McCorduck (1983) The Fifth Generation. Reading, Mass: Addison-Wesley.

Fodor, J. A. (1980) Fixation of belief and concept acquisition. In H.Piathelli-Palmanni (ed) Language and Learning: RKP.

Fodor, J. A. and Z. Pylyshyn (1981) How direct is visual perception? Cognition 9: 139-196.

Furth, M.G. (1981) Piaget and Knowledge theoretical foundations. Chicago: University Press.

Gardner, H. (1985) The Mind's New Science: a history of the Cognitive Revolution. New York: Basic Books.

Gray, J. (1987) "The Mind-Body Identity Theory". In C. Blakemore and S. Greenfield (eds) Mindwaves. Oxford: Basil Blackwell.

Gibson, J.J. (1979) An Ecological Approach to Visual Perception. Boston: Houghton Mifflin.

Hawking, S. (1988) A brief history of time. Bantam: London.

Hayes, P. (1987) "What the Frame Problem is and isn't". In Z. Pylyshyn (ed). The Robot's Dilemma. Norwood, N.J.: Ablex.

Heidegger, M. (1927) Sein und Zeit (Being and Time). Oxford: Basil Blackwell (1967)

Johnson-Laird, P.N. (1988) The Computer and the Mind: an introduction to Cognitive Science. London: Fontana.

Lighthill, J (1972) AI: a general survey. London: Science Research Council.

MacQuarrie, J. (1972) Existentialism. New York: World Publishing Co.

Merleau-Ponty, M (1962) The Phenomenology of Perception London: Routledge and Kegan Paul.

Minsky, M. and S. papent (1969) Perceptrons: An Introduction to Computational Geometry. Cambridge, Mass: MIT Press.

Minsky, M. (1986) The Society of Mind. New York: Simon and Schuster.

Morgan, M. (1979) "The Two Spaces". In N. Bolton (ed) Philosophical problems in psychology. New York: Methuen.

Neisser, U (1976) Cognition and Reality. San Francisco: W.H. Freeman.

Newell, A. and H. Simon (1972) Human Problem-solving. N.J.: Prentice-Hall.

Nirenburg, S., V. Rasking and A. Tucker (1987) "The structure of interlingua in TRANSLATOR". In S. Nirenburg (ed) Machine Translation. Cambridge, England: CUP.

Nolan, J. (1991) "The Computational metaphor and environmentalism". AI and Society (forthcoming)

Passmore, J. (1966) A hundred years of Philosophy. Middlesex: Penguin

Penrose, R (1987) "Minds, Machines and Mathematics". In C. Blakemore and S. Greenfield (eds) Mindwaves. Oxford: Basil Blackwell.

Peschl, M. (1990) "Cognition and Neural Computing - An Interdisciplinary Approach". IJCNN, pp.110-114.

Piaget, J. (1926) The Language and Thought of the Child. London: RKP.

Piaget, J. (1958) The growth of logical thinking from childhood to adolescence. London: RKP.

Piaget, J. (1970) Genetic Epistemology. New York: Columbia University Prses.

Piaget, J. (1972) Problemes de psychologie genetique. Paris: Denoel/Gauthier.

Pierce, J. (1966) "Language and Machines: Computers in Translation and Linguistics". Publication 1416, National Academy of Sciences/National Research Council.

Polanyi, M. (1958) Personal Knowledge. London: RKP.

Pylyshyn, Z. (1984) Computation and Cognition: Toward a foundaton for Cognitive Science. Cambridge, Mass: Bradford Books, MIT Press.

Pylyshyn, Z. (ed) (1987) The Robot's Dilemma: The Frame Problem in Artificial Intelligence. Norwood, N.J.: Ablex.

Rosenblatt, F. (1962) Principles of Neurodynamics. Washington, D.C.: Spartan Books.

Ryan, B. (1991) "AI's identity crisis". Byte, Jan 1991, pp.239-248.

Schank, R. and R. Abelson (1977) Scripts, Plans, Goals and Understanding. N.J.: Lawrence, Erlbaum.

Smith, N. and D. Wilson (1979) Modern Linguistics: The results of Chomsky's revolution. Middlesex: London.

Steiner, G. (1978) Heidegger. London: Fontana.

Still, A. (1979) "Perception and Representation". In N. Bolton (ed) Philosophical problems in Psychology. New York: Methuen.

Stillings, N., M. Feinstein, J. Garfield, E. Russland, D. Rosenbaum, S. Weisler, L. Baker-Ward (1987) Cognitive Science: An Introduction. Cambridge, Mass: Bradford Books, MIT Press.

Szentagothai, J. (1987) "The Brain-Mind relation: a pseudoproblem?" In C. Blakemore and S. Greenfield (eds) Mindwaves. Oxford: Basil Blackwell.

Tanimoto, S. (1990) The Elements of AI. Oxford: WH Freeman.

Thines, G. (1977) Phenomenology and the Science of Behaviour. Birkenhead: Unwin.

Thompson, H. (1983) "NLP: A critical analysis of the field". In Y. Wilks and K. Sparck-Jones (eds) Automatic NL Parsing. Chichester: Ellis-Howard.

Thompson, W. and T. Peng, (1990) "Detecting Moving Objects". International Journal of Computer Vision, 4, 39-57.

Tomita, M. (1986) Efficient Parsing for Natural Language. Kluwer: Boston.

Wetherick, N. (1979) "The foundations of psychology". In N. Bolton (ed) Philosophical problems in Psychology. New York: Methuen.

Wilks, Y. (1990) "Some comments on Smolensky and Fodor". In D. Partridge and Y. Wilks (eds) The Foundation of AI, CUP.

Winograd, T. (1972) Understanding Natural Language. New York: Academic Prses.

Winograd, T. and C.F. Flores (1986) Understanding Computers and Cognition. Norwood, N.J. : Ablex.

Winograd, T. (1990) "Thinking Machines: Can there be? Are we?" In D. Partridge and Y. Wilks (eds) The foundations of AI. England: CUP.

Wittgenstein, L (1922) Tractatus Logico-Philosophicus. Oxford: Basil Blackwell.

Wittgenstein, (1967) Philosophical Investigations. Oxford: Basil Blackwell.

Towards an Adequate Cognitive Model of Analogical Mapping

Mark T. Keane

* Department of Computer Science,
University of Dublin, Trinity College,
Dublin 2, Ireland

Stuart Duff

School of Psychology,
University of Wales College of Cardiff, Cardiff,
CF1 3YG, Wales

Abstract

This paper outlines the some of the pre-requisites of an adequate theory of analogical mapping. It reviews three computational models of analogy and considers the extent to which they approximate people's analogical behaviour. An experiment is reported which performs a competitive test between the three models, which finds results that support only one model (i.e., the Incremental Analogy Machine). The implication of this research for theories of analogical thinking and connectionist models of high-level cognition are discussed.

---

* Correspondence should be sent to this address.

1. Introduction

The central process in analogical thinking is that which draws the analogy. This process performs an analogical mapping between two domains of knowledge (see e.g., [1], [2], [5], [6], [7], [8], [9], [11], [12]). Typically, when an analogical mapping is made two distinct computations occur; first, the corresponding concepts in both domains are *matched* and, second, a portion of the conceptual structure of one domain is *transferred* (or carried over) into the other domain to form the basis of analogical inferences. For example, if you want to understand why electrons revolve around the nucleus in the atom, and are told that the atom is like a miniature solar system (see [5]), you would *match* the corresponding REVOLVES relations in both domains and *transfer* relations of ATTRACTION from the solar system domain to apply in the atom domain. In this paper, we will be predominantly concerned with the sub-process of analogical matching, with a view to developing an adequate cognitive model of the phenomenon.

1.1 Problems in Analogical Mapping

In general, when people analogically map the concepts in two domains they have to solve several tricky computational problems. To appreciate these problems consider the following mapping task from Holyoak & Thagard (see [10]):

A	B
Bill is smart.	Fido is hungry.
Bill is tall.	Blackie is friendly.
Tom is timid.	Blackie is frisky.
Tom is tall.	Rover is hungry.
Steve is smart.	Rover is friendly.

In this task, subjects are asked to say which things in list A correspond to which things in list B (ignoring the meaning of the words). Essentially, subjects have to discover a one-to-one mapping between all the individuals and attributes in list A and list B. People find this mapping problem difficult (if you want to solve it yourself, cover the next sentence). The unique one-to-one mapping which solves the problem involves matching Steve and Fido, Bill and Rover, Tom and Blackie, smart and hungry, tall and friendly and timid and frisky.

This task raises many of the difficulties involved in analogical mapping. First, for any two domains the total number of possible matches can be very large and will soon become intractable. Winston [18] pointed out that between one domain containing N1 elements and another containing N2 elements, there are N1!/(N1-N2)! possible matches. Even in a the above task, where each domain has only only six elements, there are 720 possible matches. So, it is important to reduce the number of matches that need to be considered.

Second, even though analogies usually involve isomorphic matches between both domains, many-to-one and one-to-many matches may occur and have to be disambiguated to achieve this isomorphism. For example, *smart* may match *hungry* or *friendly* or *frisky* and the correct match can only be determined by eliminating the inconsistent matches which follow from all but one of these matches.

Third, to achieve an isomorphic mapping one needs to find a consistent *set of matches* between the two domains, but there may be a large number of such match-sets to choose from. The optimal match set in our example involves all the entities in both domains but there are many others; for instance, a less optimal match-set might just involve matches between Bill & Rover and their respective attributes.

1.2 An Adequate Theory of Analogical Mapping

Palmer [17] has pointed out that any adequate theory of analogical mapping will have to operate on several levels of description (for similar ideas see [16]). At the highest level, one needs to characterise the *informational constraints* implied by the task situation; this level is concerned with describing what an analogy is, what needs to be computed to produce the appropriate outputs given certain inputs (like Marr's computational level). Below this level is the level of *behavioural constraints* which have to capture the empirical facts of people's observable analogical behaviour (Marr's algorithmic level). Hence, this level should include constraints that predict when one analogy is harder than another and the sorts of errors that people produce. Finally, there is the level of hardware constraints which aims to capture the hardware primitives of analogical thought (Marr's hardware level).

In the case of analogical mapping, analogy theorists have initial descriptions at each of the first two levels but have been silent on the last level. The current characterisation of the informational constraints on analogical mapping capture what makes a particular comparison analogical (see section 1.3). Several computational models of analogy have been proposed which implement these constraints. These models have been used to make predictions at the behavioural level, but most of these models lack significant behavioural constraints (see section 1.4). As such, they can predict idealised analogical behaviour but fail to predict errors and other aspects of performance. We will argue that there is only one model which explicitly deals with behavioural constraints; namely, the Incremental Analogy Machine ([13], [14], [15]). This model makes novel behavioural predictions which are supported by an experiment using the above attribute-mapping problem (see section 2). This work has several important implications for the modelling of analogical mapping which we will discuss later (see sections 3 and 4).

1.3 Informational Constraints on Analogical Mapping

We have already seen, that there are several problems associated with achieving optimal analogical mappings. Several informational constraints have been proposed to solve these problems (see e.g., [5], [10], [14]).

The most important set of constraints are *structural constraints*. These constraints are used to enforce a one-to-one mapping between the two domains ([3], [4], [10]). This is done using several techniques:

- *matches are made only between entities of the same type*; for example, only attributes are matched with attributes, objects with objects and two-place predicates with two-place predicates. This reduces the total number of matches that needs to be considered (see [5], [10]).
- *exploit structural consistency*; that is, if the propositions REVOLVES(A B) and REVOLVES(C D) match, then the arguments of both should also be matched appropriately, A with C and B with D. This is especially useful in eliminating many-to-one and one-to-many matches (see [3], [4]).
- *favour systematic sets of matches* (Gentner's [5] systematicity principle); this proposes that if one has two alternative sets of matches then the match-set with the most higher-order connectivity should be chosen. This aids the choice of an optimal match-set from among many match-sets.

These techniques have been shown to be very powerful. In many cases, structural constraints alone can find the optimal mapping between two domains (as in the above attribute-mapping case).

A *similarity constraint* can also be used to reduce the number of matches considered or to disambiguate between alternative matches. When this constraint is applied only identical concepts are matched between the two domains [5] or, more loosely, semantically-similar concepts are matched [7]. Semantic similarity can be used to disambiguate because if one match in a set of one-to-many matches is more similar that the others then it can be preferred.

A final constraint on analogical mapping is the *pragmatic constraint* (e.g., [9] and [11]). Again, this may disambiguate sets of matches. For example, if in a certain analogical mapping situation one match is pragmatically more important (or goal-relevant) than other alternatives then it will be preferred over these alternatives.

These three sets of constraints characterise the high-level, informational constraints on analogical mapping. They are what makes a particular comparison between two domains, an analogical comparison. While they will produce analogies that parallel human behaviour, in terms of inputs and outputs, they are not in themselves sufficient to characterise all the facets of human behaviour.

1.4 Computational Models of Analogical Mapping

The computational models of analogy proposed in the last few years have implemented the above constraints, albeit in different ways. That is, some models have used conventional, symbol processing techniques whereas others have used parallel processing techniques. Consider each in turn.

1.4.1 The Structure Mapping Engine

Falkenhainer et al.'s [3, 4] *Structure Mapping Engine* (SME) was the first model to show the importance of structural constraints (although it can also model similarity and pragmatic constraints). It operates in several distinct stages: (i) all structurally consistent matches between the items in both domains are made (e.g., all the attributes in list A are matched with those in list B and all the objects in A with those in B); (ii) each of these matches (or Match Hypotheses as they are called) is assigned a "goodness" score based on factors like the nature of the match and structural consistency; (iii) all the consistent combinations of these matches are formed to construct match-sets (or G-maps); (iv) all of these match-sets are then evaluated using criteria like their systematicity and the optimal match-set is chosen.

SME can also be used as a tool for exploring methods of comparison because it can be loaded up with different match rules. These match rules determine whether two items can be matched and also assign the goodness scores to different match-hypotheses. The standard SME rule set contains rules that only allow matches between predicates that are identical and assigns goodness scores based on structural consistency and other factors. In order to simulate the present attribute-mapping problem it is necessary to load up a set of new match rules -- which we call the *abstract rule set* -- which allow matches between non-identical predicates.

1.4.2 Analogical Constraint Mapping Engine

Another model of analogising is the *Analogical Constraint Mapping Engine* (ACME) by Holyoak & Thagard [10]. Like SME, ACME implements structural constraints but in a very different manner; it uses a localist network and parallel constraint-satisfaction. ACME forms a network of units, each of which represent a match between elements in both domains. So, for the attribute-mapping problem, the network would have a unit for the match between smart and hungry and a unit for the match between Steve and Blackie. Match units are only formed between elements of the same type (e.g., objects with objects, attributes with attributes, propositions with propositions).

All of these units are connected by weighted links, which are either excitatory or inhibitory. The pattern of weights between the units is used to implement the various constraints. Consider how structural constraints are implemented. Inhibitory weights are established between units that represent alternative matches; that is, the unit *smart=hungry* will inhibit the units *smart=frisky* and *smart=friendly*. This establishes a pressure in the network towards an isomorphic mapping. Similarly, to implement structural consistency, excitatory weights are established between units that are in the same proposition. For example, the *smart=hungry* unit will have an excitatory link to the *steve=fido* unit.

After building the network with the appropriate units and links, it is run until it stabilizes (i.e., until the units reach their asymptotes). At this point, the units with the highest activation represent the optimal set of matches between the two domains.

1.4.3 Incremental Analogy Machine

Finally, Keane & Brayshaw's [15] (also [13], [14]) *Incremental Analogy Machine* (IAM) implements all the constraints used in ACME and SME. However, unlike SME and ACME it also tries to introduce several behavioural constraints. The two main behavioural constraints are (i) working memory limitations and (ii) knowledge of the target domain. These two factors will, it is argued affect the ease of analogical mapping and the likelihood of errors. Concentrating on the working memory limitations, we have proposed that they reduce the likelihood that people consider all possible match-sets when performing an analogical mapping. People are more likely to consider only a small number of the possible match-sets.

In order to meet this behavioural constraint, we constructed a computational model that performs analogical mapping in an incremental fashion. That is, rather than generating all possible match-sets, it attempts to build just one match-set resolving ambiguities in a local fashion. It may even use fairly arbitrary means to resolve an ambiguity. For example, if a one-to-many set of matches is encountered and there are there are no grounds for preferring one match over others, IAM will simply choose the first match in the set. Clearly, there will be occasions where IAM will commit itself to one set of matches only to find that it fails. When this happens, it will undo this match-set and start constructing an alternative match-set. So, while IAM tries to avoid constructing all possible match-sets it may backtrack repeatedly to construct several alternative match-sets.

As an implementation of the informational constraints, IAM can draw analogies like SME and ACME. However, as it takes into account various behavioural constraints it should be a more accurate predictor of people's behaviour. In the next section, we consider one such prediction using the attribute-mapping task.

2. An Experiment On Ordering Effects

2.1 Predicting Behavioural Differences

As we saw in the previous section, IAM operates incrementally and may just choose the first from a set of alternatives that it encounters. In the attribute-mapping problem, there are a huge number of ambiguous matches and there are very few criteria for choosing one match over another. As such, this is an ideal situation to see whether, as IAM proposes, people make the rather arbitrary choice of the first match in a set of alternative matches. Thus, the key prediction made for the attribute-mapping problem is that under certain conditions the order in which the lists are presented will lead to behavioural differences in analogical mapping. When Holyoak & Thagard [10] experimented with the attribute-mapping problem they randomised the order of presentation of the attributes in both lists, masking any order effects in the task. We believe that order effects can be demonstrated for at least two versions of the problem, based on our assessment of what makes the attribute mapping problem difficult.

Table 1 The Two Versions of the Problem Used in the Experiment (*the singleton)

Singleton-First		*Singleton-Last*	
A	B	A	B
Steve is smart.*	Fido is hungry.*	Bill is smart.	Fido is hungry.*
Bill is tall.	Blackie is friendly.	Bill is tall.	Blackie is friendly.
Bill is smart.	Blackie is frisky.	Tom is timid.	Blackie is frisky.
Tom is tall.	Rover is hungry.	Tom is tall.	Rover is hungry.
Tom is timid.	Rover is friendly.	Steve is smart.*	Rover is friendly.

In the attribute-mapping problem, each list has two individuals (e.g., Bill and Tom) with two attributes and a remaining individual (i.e., Steve) who has just one attribute (see Table 1). This is very important because matching up the single individuals in both lists (i.e., Steve and Fido) is the key to achieving the isomorphic mapping. The presence of these single individuals with one attribute (which we will call *singletons*) disambiguates the set of matches between the two lists (this also applies to the single attribute in both lists).

Hence, if we give subjects a version of the problem in which both of the singletons are first, on the basis of IAM's predictions, the mapping should be easy because people will choose this first match and the remaining matches will fall into place. However, if we put list A's singleton last (see Table 1), subjects should not make the key match between the singletons until they have considered several alternative match-sets that fail. Hence, the mapping for the Singletons-First version should be much easier than for the Singletons-Last version. When we run these problems on IAM, this is indeed what we find. However, SME and ACME produce exactly the same outputs for both versions (see section 3 for further details on the simulations). But, let us see what people really do.

2.2 Method

2.2.1 Subjects & Design

Twenty-three undergraduates at the University of Wales College of Cardiff took part voluntarily in the experiment. The experiment had a between-subject design and subjects were assigned at random to one of the two conditions. Three subjects had to be excluded from the experiment prior to data analysis because they misunderstood the experimental instructions. Data analysis was carried out on the remaining 20 subjects, who were equally distributed across the two conditions.

2.2.2 Materials & Procedure

The materials consisted of two abstract versions of the attribute-matching problem (see Table 1). In the Singleton-First version the singletons were at the top of both lists, while in the Singleton-Last version the singleton in list A was in the last position, while the Singleton in list B was in the first position (see Table 1).

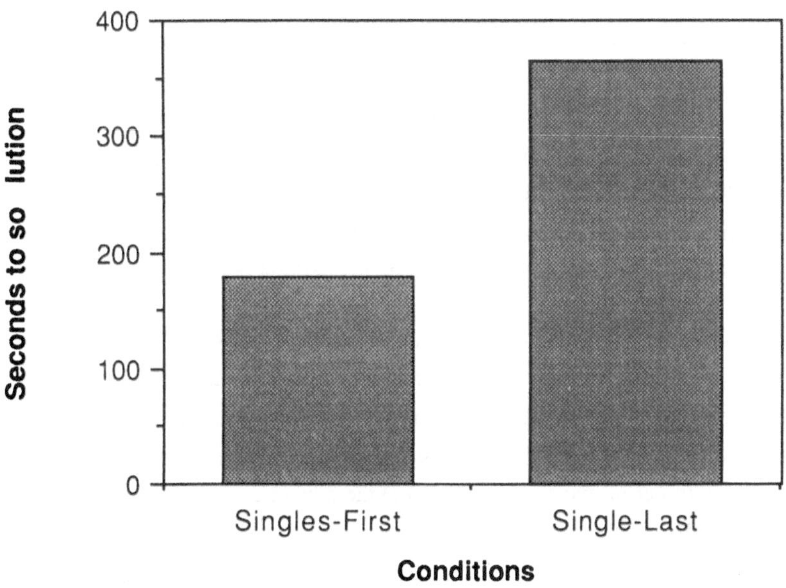

Figure 1 The Solution Times Taken in the Two Conditions of the Experiment

Subjects were instructed in writing that their "task is to figure out what in the left set corresponds to what in the right set of sentences". This sentence was followed by another which said that "The meaning of the words in the sentences is irrelevant". A single column below list A listed the names of the individuals and attributes in that list. Next to each was a space for subjects to write the corresponding name or attribute from list B. The order of remaining sentences (i.e., the non-singletons) in each list was randomised with the proviso that sentences with attributes about the same individual were kept together.

2.3 Results & Discussion

The slight change in the ordering of the singletons has a marked effect on the ease of analogical mapping (see Figure 1). Subjects in the Singleton-First condition were almost twice as fast at solving the problem (*M* = 178.0 secs) compared to the Singleton-Last condition (*M* = 363.1 secs) [Mann-Whitney $U = 7$, $p < .005$, 1-tailed].

3. A Tale of Three Models

We have seen that there are marked behavioural differences for these two versions of the attribute-mapping problem. In this section, we look at the results of the simulation work in more detail to give a better idea of the ways in which the models differ. In order to do this we need some common grounds for comparing the three programs and determining the complexity of their processing on the different versions of the problem. All the programs make roughly the same number of matches[1]. However, they differ in the number of match-sets they consider in finding the optimal mapping.

It is fairly easy to assess the number of match-sets produced in IAM and SME, but with ACME it is more difficult. Under one interpretation, ACME always only produces one match-set, namely that set of match units that have the highest activations when the network stabilizes. However, Holyoak & Thagard's index of the difficulty of a mapping is the number of cycles (or time-steps) the network takes before it stabilizes. We, therefore, believe it is more accurate to consider the state of the network on each cycle as a possible match-set. So, the total number of match-sets for an ACME run will be equal to the number of cycles it takes to stabilize. There is one further slight problem with this measure. The network can take a different number of cycles to settle into the optimal mapping depending on the initial parameters adopted. We, therefore, adopted parameter values that allowed the fastest stabilization of the network into the optimal mapping (by setting the decay parameter to 2).

---

[1] SME and IAM only consider matches between attributes and objects in both domains (about 34 in all) but ACME makes marginally more matches (i.e., 43) because it also computes matches between propositions. SME and IAM had to be given special match rules that allowed matches between non-identical predicates.

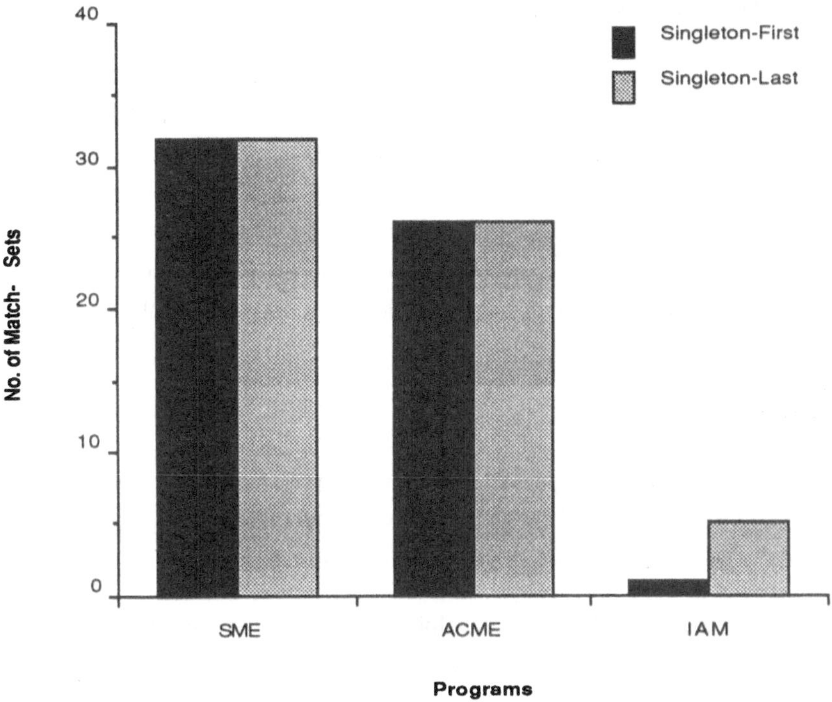

Figure 2 The Relative Number of Match-Sets Considered in the SME, IAM and ACME on the Singleton-First and Singleton Last Versions of the Problem

Using these measures, Figure 2 shows the number of match-sets considered for the Singleton-First and Singleton-Last versions of the task in each of the programs. Two notable observations can be made from this data. First, only IAM produces a difference in the number of match-sets generated for the different versions of the problem. In the Singleton-First version it only generates one match-set, the optimal one, and in the Singleton-Last version it considers five match-sets. Second, IAM is hugely more efficient in the number of match-sets it considers. SME considers 32 match-sets for these problems and ACME considers 25. In short, IAM is more efficient than the other programs and it also provides a closer approximation to subjects behaviour on the task.

4. Wider Implications

The present work was wider implications for connectionist models in cognitive science. ACME was a very important program because it showed that an example of high-level cognition, namely analogical thought, could be modelled by connectionist techniques. The present work suggests that ACME is not a wholly adequate model of analogical behaviour. It fails to capture the more serial aspects of analogising manifested in the present study. This, therefore, re-opens the question of whether the nature of high-level cognition can be captured singularly by either conventional, symbolic AI or by the new connectionist, 'brain-style' AI.

References

[1] Burstein, M.H. Analogical learning with multiple models. In T.M. Mitchell, J.G. Carbonell & R.S. Michalski (eds.), *Machine Learning: A Guide to Current Research.* Kluwer Academic Publishers, Lancaster, 1986

[2] Carbonell, J.G. Derivational analogy: A theory of reconstructive problem solving and expertise acquisition. In R.S. Michalski, J.G. Carbonell & T.M. Mitchell (Eds.), *Machine Learning II: An Artificial Intelligence Approach* . Morgan Kaufmann, Los Altos, Calif., 1986.

[3] Falkenhainer, B., Forbus, K.D., & Gentner, D. Structure-mapping engine. *Proceedings of the Annual Conference of the American Association for Artificial Intelligence*, 1986.

[4] Falkenhainer, B., Forbus, K.D., & Gentner, D. Structure-mapping engine. *Artificial Intelligence, 41,* 1-63, 1989.

[5] Gentner, D. Structure-mapping: A theoretical framework for analogy. *Cognitive Science*, *7* , 155-170, 1983.

[6] Gentner, D. Mechanisms of analogical learning. In S. Vosniadou & A. Ortony (eds.), *Similarity and Analogical Reasoning.* CUP, Cambridge, 1989.

[7] Gick, M.L., & Holyoak, K.J. Analogical problem solving. *Cognitive Psychology*, *12*, 306-355, 1980.

[8] Gick, M.L., & Holyoak, K.J. Schema induction in analogical transfer. *Cognitive Psychology*, *15*, 1-38, 1983.

[9] Holyoak, K.J. The pragmatics of analogical transfer. *The Psychology of Learning and Motivation, 19* , 59-87, 1985.

[10] Holyoak, K.J., & Thagard, P. Analogical mapping by constraint satisfaction. *Cognitive Science, 13,* 295-355, 1989.

[11] Keane, M. On drawing analogies when solving problems: A theory and test of solution generation in an analogical problem solving task. *British Journal of Psychology, 76*, 449-458, 1985.

[12] Keane, M.T. *Analogical Problem Solving.* Ellis Horwood, Chichester, (New York: Wiley), 1988a.

[13] Keane, M.T. Analogical Mechanisms. *Artificial Intelligence Review, 2(4)*, 229-251, 1988b.

[14] Keane, M.T. Incremental analogising: Theory and model. In K.Gilhooly, et al. (Eds.). *Lines of Thinking.* John Wiley, Chichester, 1990.

[15] Keane, M.T., & Brayshaw, M. The Incremental Analogical Machine: A computational model of analogy. In D. Sleeman (Ed.), *European Working Session on Machine Learning.* Pitman, London, 1988.

[16] Marr, D. *Vision.* Freeman, San Francisco, 1982.

[17] Palmer, S.E. Levels of description in theories of analogy. In S. Vosniadou & A. Ortony (Eds.), *Similarity and Analogical Reasoning.* CUP, Cambridge, 1989.

[18] Winston, P.H. Learning and reasoning by analogy. *Communications of the ACM, 23* , 689-703, 1980.

HOST: A HOlistic System Theory.

Gerard Hartnett
Humphrey Sorensen

Computer Science Department,
University College Cork, Ireland.
email: stcs8013@iruccvax.ucc.ie

Abstract

In this paper we propose an alternative theory of computation. This theory is based on interconnected interacting simple processing elements which generate complex behaviours. Our intention is to provide a new viewpoint from which we can construct, describe and use computational artifacts.

1.0 Introduction.

One of our chief concerns as software developers is the much discussed "software crisis" [37]. This crisis surfaced recently in two of the more popular computer magazines. In the September 1991 Communications of the ACM, the Inside Risks column advocated a more holistic methodology for constructing software systems. In the September 1991 issue of Byte magazine, the U.S. National Critical Technologies Panel was discussed. This panel outlined four technological areas which are of extreme importance to the future of computing and are very much in need of concentrated research work. We feel HOST can make a valuable contribution to three of these areas namely, massively parallel systems, advanced software techniques, and computer modelling/simulation.

The influences on this work are multi-disciplinary in nature. These include language/action theories [38][39], the society of minds [28], connectionism [33], distributed artificial intelligence [5], artificial life [17], theories of communication and concurrency [14][23], chaos [10], and the object orientated paradigm [6]. The common themes of these disciplines are structured interaction, intricate interconnection, and emergent complex behaviour. These disciplines will be discussed in some detail. We do not wish to take sides in the connectionist/symbolist debate - we feel that HOST is widely applicable, and will be an enabling technology in both camps.

New software technologies are becoming increasingly difficult to visualise and construct in traditional serial machine-oriented programming languages. Computing systems are now open systems [13] (in Hewitt's terminology) and are difficult to construct even when using so called human oriented methodologies such as logic programming.

HOST views a complex system as a number of simple interconnected entities called "critters". Everything in a HOST network is a critter. Critters communicate using conversations based on structured sequences of utterances or messages in a potentially non-linear fashion. Critters can be explicitly connected or can converse with others in the same environment. Critters have internal properties (contain other critters) and a number of types of conversation with which they can interact with other critters. Conversations can have a number of different states. One of the critters participating in a conversation will drive each state transition by creating an utterance. Within each state both critters can start other conversations. These conversations can be constructed using the sequence and concurrency concepts of CSP.

It is hoped that conversation based communication will provide a more natural and useful metaphor for the design of open systems through an increased awareness of commitment, structure of interaction and breakdown on the part of the system developer. HOST provides an implementation strategy for open systems, distributed AI technology, the society of minds and neural nets. While the emphasis has been on the applicability of HOST to new technologies, HOST can also be used in conventional settings. We can also see its use as a system and application integration tool.

HOST is currently being used at QC-DATA to design and implement a distributed processing environment using TCP/IP interconnected SPARCstations for resource intensive image processing work. A ContractNet type negotiational metaphor was used. Prototypes and simulations of the system have been completed, the results of which will be presented.

2.0 Theoretical Basis.

This project was initially inspired by the work of Winograd and Flores and work in implementing a parallel execution model for Prolog in Occam II [12]. The society of minds [28] was a later but equally important influence on this work. As this project is multi-disciplinary in nature, this section is devoted to introducing and explaining our interpretation of the works which influence HOST.

2.1 Foundation for Design.

In [38] Winograd and Flores expound a new foundation for the design of future computer technology. They are concerned with revealing the assumptions and blindness created by working within the "rationalistic tradition". Winograd and Flores make use of work in hermeneutics, the cognition of living systems, speech act theory and artificial intelligence, to argue their case. To conclude they propose a new type of software application called the Coordinator™.

2.1.1 Heidegger & Gadamer

Winograd and Flores cite the philosophies of Heidegger and Gadamer as important influences. Heidegger and Gadamer question the objectivist & subjectivist views of psychology.

> *"A person (subject) is not an individual subject or ego, but a manifestation of Dasein (being-in-the-world) within a space of possibilities, situated within a world and a tradition". [38]*

For Heidegger, interpretation and being in a situation are fundamental. Our being in the world interacting with other people puts us in a state of "thrownness". Our constant actions and deliberations prevent us from being able to detach ourselves from a situation.

Heidegger also introduces the terms "readiness-to-hand", "present-to-hand" and "breakdown". It is asserted that objects and properties are not inherent in the world, but arise as a result of a breakdown at which time they become present-at-hand. If an entity is ready-to-hand then no explicit recognition of the existence of the object (or its properties) takes place.

2.1.2 Cognition in Living Systems

Humberto Maturana is one of the leading physiology of vision researchers. Maturana denies there is an objectively recognisable property of an entity in an environment, or that perception is the capturing of that property. He has produced a theory of living systems based on his experiences and observations.

Maturana views the nervous system as a closed system. The organisation of a living system is autopoetic. Maturana introduces the term autopoesis to the explain the need for a system to continually transform and regenerate its component parts. This phenomenon (autopoesis) can be applied to systems existing in any domain in which components and unities can be identified. Such a system will hold constant its organisation and define its boundaries through the continuous production of its components.

Two explanations of living systems can be given. The system can be described in terms of its structure and how the structure determines behaviour. We can also describe the pattern of interactions by which the structure came to be, and the relationship of those changes to how the system acts. This is what Maturana refers to as a cognitive domain.

An organism, which is a structure-determined system that has the potential for disintegration, will participate in adaptation and evolution. Organisms of the same and different kinds can be the sources of perturbations for another organism. The interacting organisms engage in structural coupling and generate interlocking patterns of behaviour called consensual domains. Autopoesis demands and directs this structural coupling.

2.1.3 Speech Act Theory

Speech acts can be defined as meaningful acts by speakers in a shared activity. Every speech act occurs within a context between a speaker and a listener with a shared background. Speech act theory describes the structure of language acts. This theory was devised by Austin in 1962 [3]. Austin examined a class of utterances that do not refer to states in the world (felicity conditions) but which constitute acts. Promising, threatening, and naming are examples of such utterances.

A student of Austin's, John Searle, classified all speech acts as incorporating one of five illocutionary points: [34]

- Assertives.
- Directives.
- Commissives.
- Expressives.
- Declarations.

The importance of the illocutionary point is its function as a specification of the future patterns of commitment entered into by speaker and listener in a conversation.

Regularities emerge in conversation, in how speech acts are related to one another. This idea is akin to Maturana's view of the cognitive domain as being both relational and historical. An example of such regularities is given in Winograd & Flores "conversation for action". The important point is that there is an interplay of requests and commissives which are directed toward explicit cooperative action.

2.1.4 Coordination Systems.

Winograd and Flores believe software tools which function in the domain of conversation to facilitate communication will be far more useful than tools which mimic human abilities. The fundamental role of a manager is seen as the generation and maintenance of networks of conversations, not decision making. Problems are more generally seen as states of irresolution. Resolution is reached through "deliberation" among one or many agents, in a state of thrownness.

Organisations exist as networks of conversations. Breakdowns occur in conversations. To resolve breakdown further directives and requests must be created. The general structure of many of these conversations is similar - only content varies. Winograd and Flores envision a ready-to-hand tool which would operate in the domain of conversation.

Ontological design through awareness of the concepts of breakdown, presentness-at-hand, and readiness-to-hand is most important. Software should be designed with consideration for possible breakdowns, and be committed to the resolution of any breakdowns. It is impossible to anticipate all breakdowns, but awareness can only help.

Language does not describe the world but create it. A systematic domain is a formal representation of a domain created through observation of recurrent breakdown. There are systematic domains in existence for almost every profession. There are some common components in all such domains, one of which is the role of language in coordinated action.

Winograd and Flores believe that the knowledge engineering approach to AI can never produce real artificial intelligence (whatever that might be). They do not, however discount, the future potential for computer systems evolving through structural evolution, without dependence on representation. What is needed is a way to design and view computer systems as structure-determined systems.

2.2 Distributed Artificial Intelligence.

Distributed Artificial Intelligence (DAI) is concerned with knowledge and control in distributed problem-solving and multi-agent systems [5]. The main problems are those of distribution of knowledge and action in the dimensions of time, semantics, logical dependency, and spatial conception.

2.2.1 Introduction

Classical artificial intelligence theories consider one intelligent agent performing tasks. In the real world, most meaningful work is accomplished by groups of people. This realisation along with the availability of concurrent computers, and large networks of interacting computers and people, has created the environment for the development of DAI concepts and techniques.

There are two main approaches to DAI.

1. Distributed Problem Solving.

   Work in solving a problem can be accomplished by distributing knowledge among nodes. The nodes cooperate to re-construct a solution. DPS systems can distribute and cope with uncertainty in data, however there is no adaptation to changing goals.

2. Multi-Agent Systems.

   A multi-agent system comprises many autonomous intelligent agents[1]. The agents can be working toward a common goal or interacting goals. The problem is the coordination of this action. Agents must reason about the nature of their interactions with other agents, as well as the nature of the problem.

2.2.2 Main areas of research.

Most distributed artificial intelligence research can be classified as being in one of four different areas. HOST can be classified as being in the communication and interaction

1. The term Agent is much used in recent AI literature and is defined by [5] as: A computational process with single/multiple locus of control and/or "intention".

section, although it does provide a framework for dealing with problems in the other areas.

- Description, decomposition and allocation of tasks.

  HOST will be applied to a task allocation problem in section 4.0.
- Communication and interaction.

  Interaction is dependant on the coordinated action of at least two agents. It is considered A communicates with B when A acts on B with linguistic actions and B reacts to A. Of particular interest are the "dialog games" of Levin and Moore [18], and the "plan based speech act theory" of Cohen and Perrault [9]. Dialog games refer to two way interactions, their initiation, and their control using partially ordered sub-goals. Dialog games are related to the conversation structures of critters. The plan based speech act theory relates belief, and intention to plans which generate speech acts. Speech acts are seen as operators of a plan.
- Coherence & Coordination.

  Coherence is related to the qualities of the entire system, such as efficiency and solution quality. Coordination refers to the degree of required synchronisation and alignment of tasks between agents. Cooperation is seen as a special case of coordination. Achieving global coherence and coordination in a system with local control is extremely difficult.
- Reasoning about plans and actions of other agents.

  An agent must have some indication of the possible effects of an action on another agent. Different levels of agent knowledge can be supplied and used. Coherence and coordination are highly dependant on the quality and use of agent's plans and actions.

2.3 Society of Minds.

The "The Society of Minds" [28] is a speculative theory on the high level functions of the human mind. Proposed by Marvin Minsky of MIT, it is based on earlier work in frame theory [25], the society theory of thinking [26], and K-lines [27]. The theory views the mind as an organised society of intercommunicating agents.

An agent has no intelligence of its own, it is through interactions in large numbers that they exhibit complex behaviour. Agents are seen as the fundamental particles of the mind. A group of agents when looked on as a unit are called an agency.

Agents are built up into vast bureaucracies, which can be considered almost hierarchical. Internal conflict in an agency will weaken its reputation with higher level agencies. Only high level agents can partake in complex communication such as negotiation. Low-level agents have simple rules for communication and can only communicate with a few other types of agents.

Agencies are not exactly hierarchies, but hetrarchies. The organisational metaphor of a hierarchy is useful but cannot be generally applied. There can be loops in the structure where an agency can communicate with a higher level agency which may have indirectly invoked it. This form of circular causality is similar to the programming concept of recursion.

Agents do not explicitly communicate information but set up the context in which other agents function. High level agents cannot communicate with low level ones directly.

[28] is made up of 270 essays which interact like agents, and creates a new context for thinking about the mind. The "society of minds" has been used in part in implementing artificial life systems [8][36].

2.4 Artificial Life.

Artificial Life (AL) has existed as a coherent discipline since the first workshop on the synthesis and simulation of living systems [17].

> *"Artificial Life involves the realisation of lifelike behaviour on the part of man-made systems consisting of populations of semi-autonomous entities whose local interactions with one another are governed by a set of simple rules. Such systems contain no rules for the behaviour of the population at the global level, and the often complex, high-level dynamics and structures observed are emergent properties, which develop over time from out of all the local interactions among low-level primitives by a process highly reminiscent of embryological development, in which local hierarchies of higher-order structures develop and compete with one another for support among the low-level entities. These emergent structures play a vital role in organising the behaviour of the lowest-level entities by establishing the context within which those entities invoke their local rules and, as a consequence, these structures may evolve in time." [17]*

2.4.1 Introduction.

Biology is concerned with the analysis of existing carbon-based life, while Artificial Life is concerned with the synthesis of possible life. Biology is difficult because general theories must be formulated from few examples. Artificial Life focuses on the properties of the organisation of matter not the matter itself. Biology can be considered a top-down science, whereas AL is bottom-up. Populations of simple machines are constructed, and interact in a non-linear way to produce emergent life-like behaviour.

The computer is the synthesis tool for AL applications. Conventional computer programming languages however are not the ideal platform for implementing highly-parallel AL models. A typical AL model has a population of programs, no global control program, each program reacts to local events, there are no global rules for interaction, and higher level behaviour is emergent.

2.4.2 GENOTYPES and PHENOTYPES.

At the core of this new discipline are the biological concepts of genotype and phenotype. Genotypes and phenotypes are terms used to describe the information explicit in DNA. Genotypes specify biological machinery while phenotypes are the behaviour that machinery generates. They can be generalised and applied to non-biological domains. A GTYPE is a generalised genotype and describes the reactions of an entity to local interactions. The PTYPE emerges out of the collected non-linear local interactions of these entities. There are many observable levels of PTYPES. There is a PTYPE for, each GTYPE instruction, for each individual entity and for the society of entities.

2.4.3 Evolution.

Trial and error is the only method by which GTYPES or PTYPES can be found. Evolution is the only general form of trial and error processing discovered so far. Natural selection is the mechanism by which evolution occurs. PTYPES are generated from an initial set of GTYPES. They interact with each other and the environment. The GTYPES of relatively successful PTYPES are copied and slightly modified. This process is repeated until there are no GTYPES remaining. Genetic algorithms are an application of natural selection to machine learning.

2.4.4 Linear and Non-linear Systems.

The simplest distinction to be made between linear and non-linear systems, is that the behaviour of a linear system is equal to the sum of the behaviours of its components. Hence, non-linear systems don't follow the rules of superposition. It is difficult to isolate the components of a non-linear system. The most interesting behaviour is a result of the interactions between the components. Analysis can be applied successfully to linear systems, whereas synthesis is the only method for the creation of desired non-linear systems.

2.5 Other Influences.

This work has also been influenced by other disciplines too numerous to expand on here. These include object-orientation [6][22], chaos theory [31][10], connectionism [33], communicating sequential processes [14], and the calculus of communicating systems [23].

3.0 Functional Description.

Critters are the fundamental components of a HOST network. The topography of a simple HOST network is shown in FIGURE 1. Each of the circled letters is a critter. A number of critters exist in an environment. Critters can explicitly connect to other critters in the same environment through bidirectional communication channels. The environment acts as a boundary of interaction. Each critter is part of a species, signified by the labels in the circles of FIGURE 1. There can be more than one instance of each critter species.

FIGURE 1 Topography of a simple HOST network.

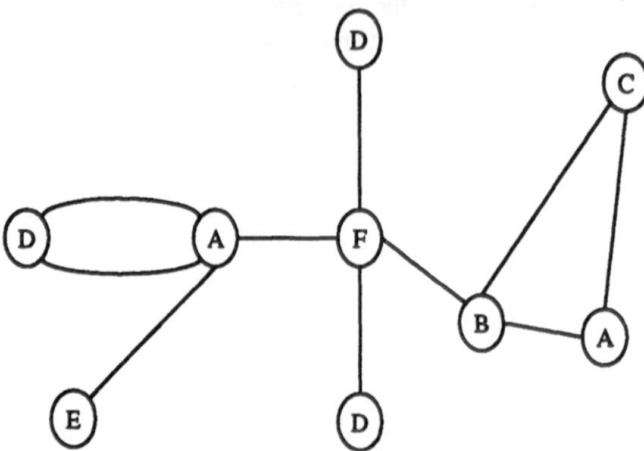

This network is essentially a hetrarchical network. If you consider the critter belonging to the F species as being the highest-level critter, you will observe a loop in the hetrarchy between B and C. This is allowed in HOST, as B will be able to start many different conversations with both A and C. There can be multiple instances of a critter species - there are two instances of both the D and A species. A critter can have more than one conversation in progress with another critter, as seen in the two conversations between D and A on the left of the diagram.

FIGURE 2 Conversations between critters.

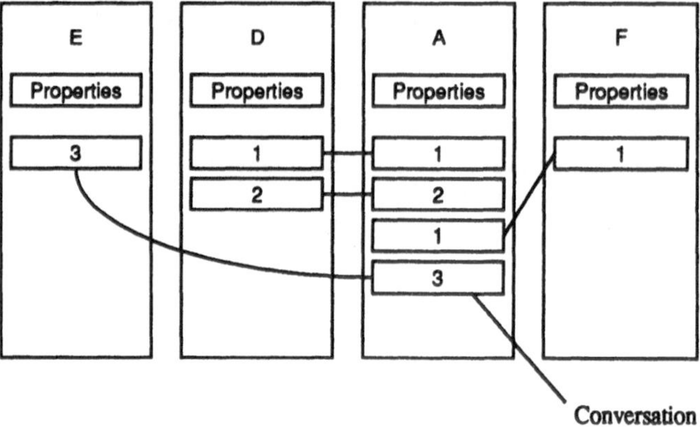

In FIGURE 2 we see a diagram of the four critters on the left of FIGURE 1. The critters are illustrated in a little more detail. Each critter has a set of properties, in the same way objects in an object orientated language have instance variables. These properties are made up of other critters. The conversation boxes contain a number which is the conversation structure identifier.

The important component of this theory is the use of "conversation" (structural coupling) as a metaphor for critter communication. The fundamental components of each conversation are utterances (speech acts). Each critter has a set of "conversation structures". A conversation structure defines the interplay of utterances which make up a type of interaction. These are specified through recognition by a designer of recurrent conversation patterns, similar to those observed in systematic domains (section 2.1.4). Each utterance in a conversation is a speech act (section 2.1.3) with illocutionary point and force. Critters create and direct utterances. Each critter can engage in more than one conversation, with more than one other critter at a time, providing more than one locus of control.

FIGURE 3 The conversation for action.

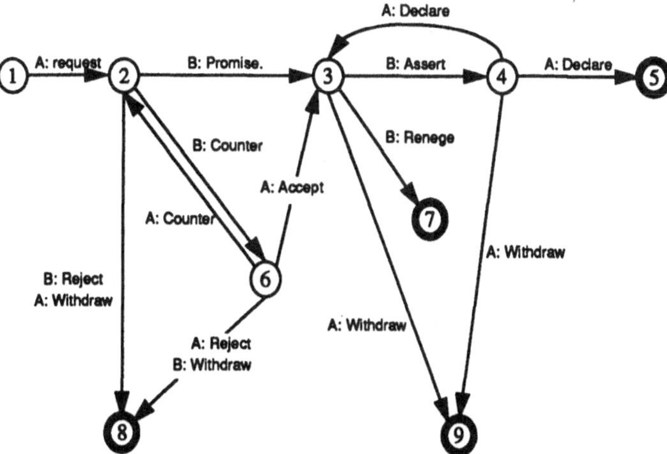

FIGURE 3 illustrates a conversation structure. A conversation is made up of a number of states, denoted as circled numbers. The heavy circles signify states of conversation completion. Utterances drive the conversation from state to state. Utterances are specified as speech acts (example section 4.2). The states of a conversation provide a local context for further conversation, while the properties provide a pan-conversational context.

Each participant in a conversation will be in the same state at the same time. Both critters will use the same conversation structure. Each critter however will perform different actions within each state. When a critter enters a state it must have some method of deciding which utterance if any to generate. This is done by executing a plan. A plan is a sequence of invocations of other sub-conversations with other critters (possibly the critters which make up the critter's properties). The results of these sub-conversations are then used to decide on the utterance and content to be generated in the conversation. These plans have the concurrency semantics of CSP and Occam. FIGURE 4 illustrates a plan which conversant B in FIGURE 3 could execute.

When the plan is begun we see a PAR node (FIGURE 4). This creates two parallel strands in the plan. In one of these strands C3 is started, in the other C1. If C3 terminates it will return its resultant state to this plan. A decision can then be made based on this resultant state to either generate the reject speech act in the conversation or do nothing. Likewise, the resultant state of C1 determines whether the counter utterance is generated or the C2 conversation is started.

FIGURE 4 Plan for conversant B state 2.

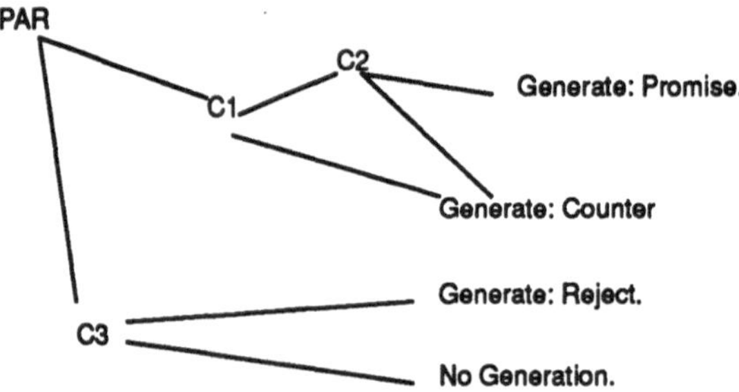

As already stated, each critter shares an environment with other critters. A critter can start a conversation with any other critter in the environment. A critter can examine the conversation structures available to each critter to build a rudimentary model of their abilities. A critter can also examine the states of conversations other critters are in to build up simple models of their intentions and beliefs.

4.0 Application & Implementation.

The principles of HOST are being applied at QC-DATA to provide tools for the analysis and design of an environment suitable for distributing large-grain resource intensive programs to loosely coupled processors such as ethernet interconnected SUN SPARCstations. This application is similar to Contract-Net [5] and also uses negotiation as a task allocation metaphor.

4.1 Outline.

"Critters" will facilitate the distributed parallel processing of jobs made up of primitive units (processes) on machines also running critters. These "critters" will be connected through communications channels implemented using TCP sockets and C++ task queues. FIGURE 5 shows the top level entities of this problem. These are encapsulated as critter species. This is a sketch outline of these species and their interactions - their interactions will be examined more in section 4.2.

- Job.

  The job critter represents a production job requiring processing. A job is essentially a sequence of programs which need to be run. Work is distributed at the individual program level. The job critter can interact with machine critters, appealing for processing time and selecting the most suitable machine for processing, considering the received replies.

- User.

  The user critter is a piece of software through which the production user interfaces with the system. The user critter interacts only with jobs. The user initiates and can examine the status of any job critter.

- Machine.

  The machine critter is essentially the interface to each machine. Its most important conversations are with job critters. These conversations involve requests for processing, declarations on ability to process, commitments to process and inquiries on the state of processing. The machine interacts on a lower level with process critters.

- Process.

  The process critter interacts only with a machine. The process critter replies to requests for estimates about resource needs, directly spawns UNIX processes, and informs the machine about the success or failure of the spawned processes.

- Administrator.

  The administrator critter is the interface point for system administration. Through the administrator, a person can monitor and control jobs, and machines.

FIGURE 5 Topology of the QC-DATA HOST network.

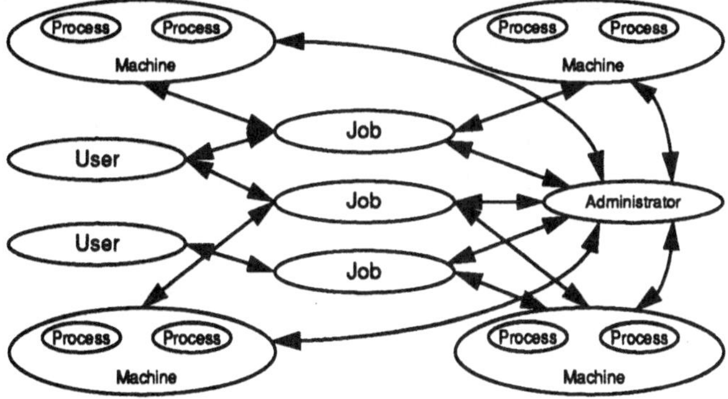

4.2 Recurrent Conversations.

Most of the conversations are fairly simple. The most interesting type of conversation however is the one between the machine and the job which is quite complex. The conversation structure is illustrated in FIGURE 6.

FIGURE 6 Request for Processing (conversation structure diagram).

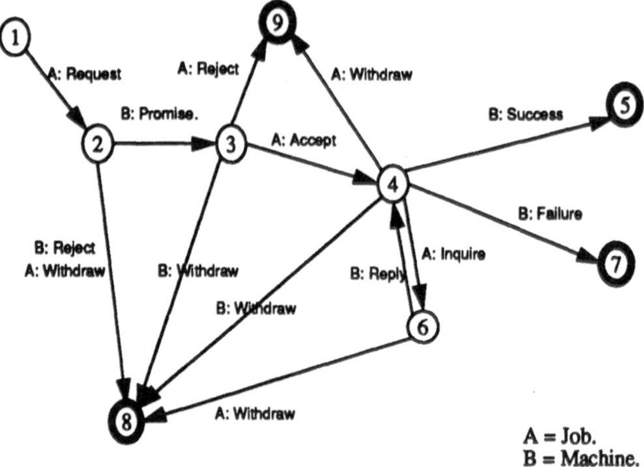

The different states of the interaction are as follows.

1. The job first requests the machine to execute the process required. It relates its priority and indicates the size of input files which would influence processing time and resource usage.
2. The machine can (following interaction with the process "critter" inside its plan for state 2) either promise to fulfill or reject the request. If the machine rejects we go to stage 8 which means the job has to find an alternative machine to run its process. With a promise to process, the machine commits to being done at a certain time.
3. The machine can withdraw this promise (as might happen if a higher priority job came to it before it moved from this state). The job will either accept or reject the commitment on the part of the machine. If it rejects we go to stage 9 which means the job must have found another more suitable machine to process on. If the job accepts we move to stage 4.
4. State 4 is where the machine is processing the job's request while the job waits. The machine can withdraw at any time forcing the job to find alternative processing arrangements. The job can poll the machine with inquiries. The polling interval will be related to the committed completion time from stage 3. If the machine cannot respond for any reason the job moves to stage 8 in the absence of a reply. The

machine can either fail or succeed to process the job's request. On success it goes to stage 5 on failure it goes to 7.

5. State 5 will trigger a message to the job, which will move the job to process another program.
6. This is where the job is waiting for a reply from the machine to its polling.
7. State 7 will trigger a message, perhaps to the system administrator, who will then investigate the apparent failure of a process.
8. The job has to find another processor for the process.
9. The interaction is finished no processing was needed by the job of the conversant machine.

This conversation will be part of a higher level conversation within the job critter, state 5 will trigger an utterance to this higher level conversation which will move the job to process another part of the job.

4.3 Results of simulations.

Tests were run on the currently in-place scheduling system. The results of the processing requirements of this test were used along with a simulator for the HOST based scheduling system. A simulation of this scheduler has been coded in C++ and test data about individual machine performance was put into this simulator. The simulator uses a time-warp like clock object. While the scheduler itself will have concurrent tasks and use the C++ task library, the simulator posted notify requests to the time-warp clock to emulate concurrency. The network was very well balanced with no machine more than 0.4% inactive (unlike the current system). It was found that eight test jobs processed in 56% of the time it took with the old system, using the same machines and programs.

The proposed system has a high degree of robustness. This would normally be difficult to achieve but becomes easier when one concentrates on the fact that the computing entities converse, and those conversations can breakdown. There will be increased flexibility on two levels. The whole topology of conversations can be changed easily at a very high level to change the behaviour of the system considerably. There is information and behaviour hiding, in that the machine process can decide what strategy to use to accomplish its work so long as it obeys the protocol of the conversation. The system has the potential to learn. The process critters accumulate records of resource needs of particular programs. When the system is in operation for a while the time estimations given by the machine to the job critter will become more accurate. In a short space of time the whole network of computers would *"evolve"* into an optimally configured distributed processing system which would adapt to changing loads and changing programs.

5.0 Conclusion.

The fundamental theme in HOST is the bottom up specification of complex behaviour using simple processing elements.

It is largely true to the foundations for design expounded by Winograd and Flores. HOST systems incorporate structural plasticity. Critters can to some degree interpret utterances. Commitment is facilitated in the interactions between critters.

HOST will be useful for solving DAI problems. HOST is at the same level of abstraction as Hewitt's open systems. The addition of "conversation" however makes critters more powerful than actors. HOST concentrates on the interaction problems of DAI but provides tools with which to manage coordination, reasoning about other agents, and conflict resolution.

HOST could be used to build a "society of minds". High level agents could be implemented directly as critters. Critters with very simple conversation structures could be used as low level agents or indeed the nodes of a neural net.

In the realm of general programming we believe "conversation" and ontological design may be a most fruitful metaphor for programming. Conversation is a fundamental and most accessible component of human life, and will promote an awareness toward possible breakdown. HOST is in effect a generalisation of object-oriented techniques. This point is best illustrated by a quote from "Understanding Computers and Cognition".

> *"Even at the simple level of providing the initial possibilities of 'make request' and 'make promise' instead of 'send message', it continually reminds one of the commitment that is the basis for language. As one works successfully in this domain, the world begins to be understood in these terms, in settings far away from the computer devices." [38]*

In the real world, computing is about interacting components configured in large complex systems. The description of such systems is not easy from the rationalistic viewpoint. There has been much interest in the application of the object paradigm to the design of operating systems and computer systems [29][30]. This idea has been validated by the proposals of Apple/IBM for an object based operating system called Pink. We feel the principles of HOST could be incorporated in these ideas, to considerably enhance such systems.

HOST is still in the early stages of conceptualisation and development. Object orientated concepts such as inheritance must be considered, intentionality and belief must be looked at more closely, and formal theories of concurrency must be incorporated in a more rigorous fashion.

6.0 References.

[1] Adams, D. Dirk Gently's Holistic Detective Agency. Pocket Books, New York, 1987.

[2] Agha, G. and Hewitt, C. Concurrent Programming using Actors: Exploiting large-scale parallelism. in [5].

[3] Austin, J.L., How to Do Things with Words. Harvard University Press, 1962.

[4] Betz, D. An Adventure Authoring System. *Byte Magazine*, May 1987.

[5] Bond, A. and Gasser, L. (eds.). Readings in Distributed Artificial Intelligence. Morgan Kaufmann Publishers, 1988.

[6] Booch, G. Object Orientated Design with Applications. Benjamin/Cummings, 1991.

[7] Brand, S. The Media Lab. Penguin Books. 1988.

[8] Coderre, B. Modelling Behavior in Petworld. in [17].

[9] Cohen, J. and Perrault, C. Elements of a Plan-Based Theory of Speech Acts. in [5].

[10] Gleick, J. Chaos. Penguin, 1987.

[11] Greif, I. (ed.). Computer Supported Cooperative Work. Morgan Kaufmann, 1988.

[12] Hartnett, G. The Design and Implementation of a Logic Programming Language for a Parallel Computer. Undergraduate Thesis, Dept. of Electronic Engineering, University of Limerick, Ireland, 1989.

[13] Hewitt, C. Offices are Open Systems. in [5].

[14] Hoare, C.A.R. Communicating Sequential Processes. Prentice-Hall, 1985.

[15] Kay, A. Microelectronics and the Personal Computer, in Peterson, G.E. Object Orientated Computing, volume 1. IEEE Computer Society Press, 1987.

[16] Kirkpatrick, S. G., Gelatt, C. D., and Vecchi, M. P. Optimisation by simulated annealing. *Science* 220, 4598, 1983.

[17] Langton, C. Artificial Life. Addison-Wesley, 1989.

[18] Levin, J. and Moore, D. Dialogue-Games: Metacommunication Structures for Natural Language Interaction. in [5].

[19] Looby, B. An implementation of simulated annealing on a transputer network. Undergraduate Thesis, Dept. of Electronic Engineering, University of Limerick, Ireland, 1989.

[20] Maturana, H. R. Biology of Cognition. 1970, re-printed in [21].

[21] Maturana, H. R., and Valera, F. Autopoesis and Cognition: The realisation of the Living. Reidel, 1980.

[22] Meyer, B. Object-orientated Software Construction. Prentice-Hall, 1988.

[23] Milner, R. Communication and Concurrency. Prentice-Hall, 1989.

[24] Minsky, M. and Papert, S. Perceptrons. MIT Press, 1969.

[25] Minsky, M. A framework for representing knowledge. in Winston, P. (ed.). The Psychology of Computer Vision. McGraw Hill, 1975.

[26] Minsky, M. The Society Theory of Thinking. in Winston, P. and Brown, R. (eds.). Artificial Intelligence: An MIT Perspective. MIT Press, 1979.

[27] Minsky, M. K-Lines: a theory of memory. in Norman, D. Perspectives on Cognitive Science. Ablex, 1981.

[28] Minsky, M. The Society of Mind. Simon & Schuster, 1987.

[29] O'Mullane, W. The Object Computer. AbaKus, Issue 2, Computer Science Department, University College Cork, Ireland, 1991.

[30] O'Mullane, W. Odyssey: Progress and Research Directions. Internal Report, Computer Science Department, University College Cork, Ireland, 1991.

[31] Pagels, H. The Dreams of Reason. Bantam Books, 1988.

[32] Peterson, G.E. Object Orientated Computing, vols 1 & 2. IEEE Computer Society Press, 1987.

[33] Rumelhart, D. & McClelland, J. (eds). Parallel Distributed Processing. MIT Press, 1986.

[34] Searle, J.R. Speech Acts. Harvard University Press, 1969.

[35] Shapiro, S. (ed.). The Encyclopedia of Artificial Intelligence. John Wiley & Sons, 1990.

[36] Travers, M. Animal Construction Kits. in [17].

[37] Winograd, T. Beyond Programming Languages, *Communications of the ACM,* 22:7 1987.

[38] Winograd, T. and Flores, W. Understanding Computers and Cognition: A Foundation for Design. Ablex Corporation, 1987.

[39] Winograd, T. A Language/Action Perspective on the Design of Co-operative Work. in [11].

Section 2:

Natural Language – Applications

THE COMPUTER AS A TUTORIAL ASSISTANT FOR THE TEACHING OF INFLEXION IN IRISH WORDS

Caoimhín MacGallóglaigh, Design Automation Co. Ltd.,Tokyo 180
Pádraig MacIonnrachtaigh,Gearóid Ó Néill, Ollscoil Luimnigh, Luimneach
Máire Ni Mhurchú,New York, Pádraig Ó Flaithearta, AIB,Dublin 4

Abstract

This paper describes a system for the teaching of inflexion of Irish words. The system attempts to provide a facility akin to a tutor - it provides rules for inflexion, examples, exercises and explanations, as well as pronunciation for each word, the meaning of words and the translation of words. The system attempts to let the user be in charge and decide, as far as possible, how to progress through the system and what facilities to use.

Introduction

The computer can be used, we feel, to assist in many aspects of language learning, particularly in relation to the basic aspects of the written form of a language, such as inflexion. A computer system allows the linguist to practice the inflexions as much as he or she feels the need to and, of course, without getting impatient. In this system, we have attempted to provide an informal tutorial environment [2], where the student feels in control. A tutor is able to guide on the main aspects of a subject and on side issues. Since it is an informal tutorial, the student does not have to follow a predetermined path nor answer a question before moving on to new material. Function keys allow the student to access various features without having to quit the current activity, just as a tutor can answer questions in ancillary areas. These related features currently include explanation facilities and information on pronunciation and meanings of words.

The CALL

The knowledge-base currently includes nouns, verbs, adjectives and prepositions. There are introductory sections, examples and exercises. In the introductory section, the rules for the inflexion are given. There are two kinds of exercise, one is to give a form or forms of a word, the other to provide the form in a sentence. The linguist selects which topic to pursue.

As when a tutor is explaining a topic, the linguist may ask about a point already dealt with or about a point about to be dealt with. In the system this is currently achieved by a facility for reading ahead or reading back from the current page,whilst retaining the current page.

In the examples and the exercises on the individual word, the screen is arranged into three windows, with an options line along the bottom of the screen. One window has the explanations (rules) for achieving the inflexion, another window presents a word for inflecting. A third window for further dialogue.

There are options to get a guide to the pronunciation in the form in quasi-IPA symbols and to get the meaning of a word[1] or the translation of the word. The system is currently restricted to the "official" forms. The IPA rendition is based on Lárchanúint don Ghaeilge [5].

The linguist choses the lesson, be it for the conjugation and the tense of a verb or the case for a noun. The word to work on is randomly selected by the system or supplied by the linguist. The rules window can be turned off.

Verbs

There are two conjugations in Irish and about twelve irregular verbs. It is from the second person imperative that the other parts of the regular verbs are formed. The tenses achieved by inflexion on the stem are the continuous present, the future, the conditional, the past, the past habitual, the present subjunctive and the imperative. Often the first person singular or plural has a "synthetic" form. The other persons usually have the same ending but use the personal pronoun explicitly.

The present tense of the first conjugation is formed as follows

first person singular	synthetic form If slender ending then append im if broad ending then append aim
second person singular	if slender append eann if broad append ann pronoun tú
third person singular	if slender append eann if broad append ann pronoun sé or sí
first person plural	synthetic form if slender append imid if broad append aimid
second person plural	if slender append eann if broad append ann pronoun sibh
third person plural	if slender append eann if broad append ann pronoun siad

cas turn	rith run
casaim	rithim
casann tú	ritheann tú
casann sé	ritheann sé
casann sí	ritheann sí
casaimid	rithimid
casann sibh	ritheann sibh
casann siad	ritheann siad

Past Tense

Verbs beginning with consonants are lenited, if possible. Verbs beginning with vowels or fh, as a result of lenition, are preceded by d'.

cas turn	ól drink
chas mé	d'ól mé
chas tú	d'ól tú
chas sé	d'ól
chas sí	d'ól sí
chasamair	d'ólamar
chas sibh	d'ól sibh
chas siad	d'ól siad

(NB the prefix d is from the particle do, which used to be placed before a verb in the past tense.)

As shown above, the verb stem may be modified at the beginning as well as at the end. Several tenses have lenition. The verb system determines first any changes to the beginning of the stem. Once the beginning is determined, the endings are obtained. The verbs are characterized by their ending in the lemma. Included in such lemma endings are áil, éigh, ígh, áil. The present tense, for example, of verbs with the stem ending in áil is formed by omitting the i and then using the rules shown above for the first conjugation present tense.

Figure 1 shows an example for the verb **cas**. The student has entered the replies and given the wrong form of the first person singular.

```
+--------------------------- The First Conjugation----------------------------+
+------------- Explanation ------------++-------------- Exercises -------------+
| To  get  the  persons  one makes the ||                                      |
| following changes , which depend  on || Verb :      cas                      |
| the verb being Broad or Slender(F7). ||                                      |
|                                      || Verbal Noun : casadh                 |
|  dún -->   dúnaim                    ||                                      |
|            dúnann tú                 ||                                      |
|            dúnann sé                 || Verbal Adjective :  casta            |
|            dúnaimid                  ||                                      |
|            dúnann sibh               ||                                      |
|            dúnann siad               ||  Singular                            |
|                                      ||                                      |
|          The 1st PERSON sing. is wrong.    casim                            |
|                                      || 2. casann tú                         |
|       ( F1 - Detailed Rules )        || 3. casann sé                         |
+--------------------------------------+|                                      |
+-------------- Messages --------------+|  Plural                              |
|cas                                   ||                                      |
|    ends in a broad consonant         || 1. casaimid                          |
|to form 1st person add im             || 2. casann sibh                       |
|but preserve leathan le leathan rule  || 3. casann siad                       |
|so add a to 2nd imperative singular   ||                                      |
|cas -> casa -> casaim                 ||                                      |
+--------------------------------------++--------------------------------------+
 F10:End  F4 How to form correct answer F5:Retry  F6:Correct Answer  Esc:Exit
```

Figure 1

Nouns

The basic part of the noun is the nominative singular and form this the other parts of the noun are formed. Generally the nouns are classified into five declensions and a group of irregular nouns.

The changes to nouns include affixes and internal changes. The first declension is the declension with the largest number of words.. There are several cases in the declensions but the main cases are the nominative and the genitive. The nominative singular 1st declension nouns end in a "broad" consonant, i.e. the ending consonantal group is preceded by a broad vowel - a o or u. The genitive singular is achieved by making the ending consonantal group slender. To render the ending slender, the letter i is inserted before the consonants (with a possibly new set of consonants)

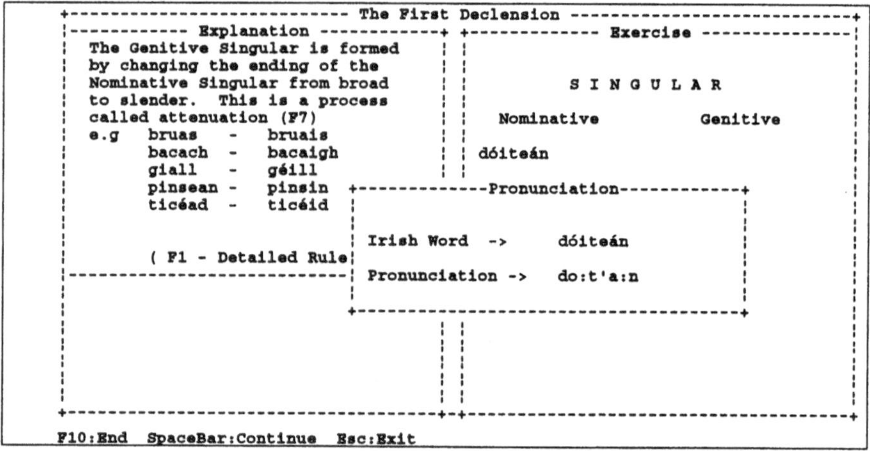

Figure 2

Figure 2 shows a session on the first declension, genitive case. The linguist has requested the pronunciation of the word, for which he or she is to provide the genitive singular.

Prepositions

Some prepositions in Irish coalesce with the pronoun, for example **ag mé** (at me) becomes **agam**. Similarly for each person. For **ag** (at) we have,

agam agat aige aice againn agaibh acu.

The prepositional pronouns also have an emphatic form. Some prepositions lead to mutation of the following noun (sometimes referred to as the dative case). The basic exercises here are to provide the form for the particular person or each of the possible forms - either normal or emphatic. There is also an exercise on phrases involving preposition, article and noun.

Mutation

Mutation in Irish takes two forms, lenition and eclipses. Lenition is usually represented in writing by the insertion of the letter h after the consonant to be lenited, for example, dhún me - past tense of dún (shut). Eclipses is shown by "eclipsing" the preceding a letter with one or two other letters, as, for example,ar an mbád - on the boat. The systems deals only with mutation in the context of the particular part of speech under consideration.

Auxiliary Facilities

In addition to the main windows, there are auxiliary windows for pronunciation, meaning and translation. In the exercises on inflexion in isolation it is possible to get the sound, in quasi-IPA symbols, [Figure 2]. The sound representation is based on Larchanuint na Gaeilge. Consonants generally represent one of two sounds, one broad one slender. The quality of the consonant is generally determined by accompanying vowel or vowels. The vowels a o u are designated broad, e i slender. IPA symbols are "approximated" following Ó Baoill and then approximated again to get similar symbols using the extended ANSI code. The word bád (boat) is shown as ba:d where b represents the IPA symbol ƀ and d represents the IPA symbol đ and a: represents long a. Bawd would be a close English transcription. It does not represent a unique vowel sound[5]. The IPA representation can be obtained for a word on which the current exercise is taking place or by typing in a word. Below is (tiny) part of the rules for pronunciation. The example shows some of the basic pronunciation representations of letter combinations with the quasi-IPA symbols. The IPA string is in quotes.

(ϵ is used to represent the neutral vowel)

```
consonant(lg,"lϵg").          consonant(mh,v).
consonant(lm,"lϵm")           consonant(lb,"lϵb").
consonant(rm,"rϵm").          consonant(rg,"rϵg").
consonant(nbh,"nϵv").         consonant(rch,"rϵx").
consonant(nm,"n'ϵm'")         consonant(nb,"nϵb").
consonant(nmh,"nϵv").         consonant(rm,"rϵm").
consonant(ch,x).              consonant(rbh,"rϵbh").
consonant(c,k).               consonant(bh,v).
```

Similarly, the meaning[1] or translation can be obtained for a word. The system keeps track of the attempts and number of errors. The error tracking is only rudimentary in the system. The initial effort has been to generate the tutorials. The linguist also has a notepad option and may exit to the DOS system and return to the "tutor".

Exercise through sentences

In addition to exercises on individual words in isolation, there is an exercise available for replacing the basic form of a word by its appropriate form in a sentence. the basic form of the word is presented in the appropriate place in a sentence. The linguist is asked to replace the word with the suitable form. These exercises are generated either by the teacher or, in conjunction with the teacher by the system semi-automatically from texts. Through a special sentence exercise generation phase, the system selects instances of words and the teacher confirms the choice.

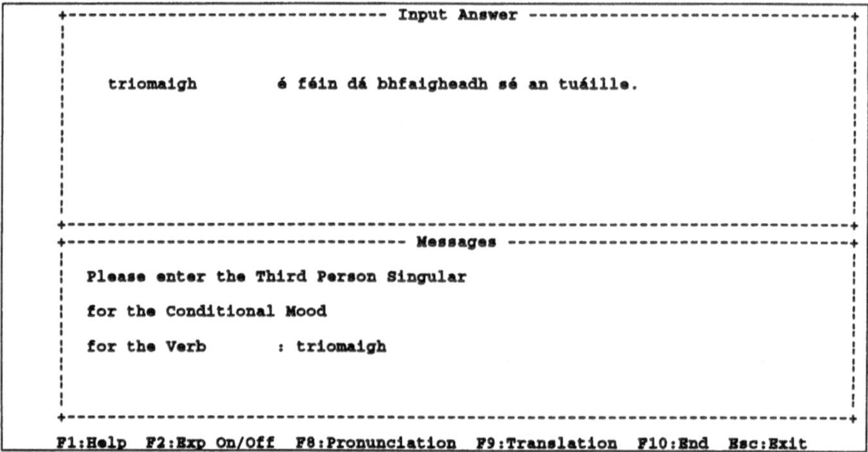

Figure 3

Current state of system

The system is written in Turbo-Prolog for IBM-compatibles. The system is in development with most of the work done on nouns, verbs and prepositions. The features mentioned have been achieved in one or more parts of the system. The words and the dialogue are held in databases. Any language can thus be supplied (provided a kindly humanoid does the translation) to be used as the intermediate language for teaching of the Irish. Currently the Irish and the English is available. Each section has explanations on how to generate the various forms from the lemma. The IBM-compatibles were chosen because they were a "minimum" machine for achieving such a system and the schools are increasingly using them. Some minor testing has been done from the point of view of ease of use. Children, from as youmg as eight years of age, and adults have tried the system and found it easy to use. Some of the ideas on the organisation of the windows were developed from cal for lexical, syntactic and code optimization [2,3,4]

Acknowledgements

I should like to thank

Máire Ni Cheilleachair, Liam Ó Laoire and Liam Ó Dochartaigh, Ollscoil Luimnigh

Dónal Ó Baoill, ITÉ, BÁC

Ciarán Ó Duibhinn, QUB, Béil Féirste

for assistance and material; families and friends for testing the system.

References and Bibliography

[1] Foclóir 3 don aos óg
An Roinn Oideachais, BÁC

[2] N. Martyn.
KBS - CAL Code Optimization Techniques, 1988
Tionscadal Céime, Ollscoil Luimnigh

[3] G.Morrisroe.
CAL for Syntax Analysis, 1988
Tionscadal Céime, Ollscoil Luimnigh

[4] M. Nolan.
CAL for Lexical Analysis, 1988
Tionscadal Céime, Ollscoil Luimnigh

[5] Foclóir Póca, 1986
An Gúm,
44 Sráid Uí Chonaill Uachtarach, BÁC 1

[6] P. Ó Flaithearta,
Córas Cainte Síntéiseach Don Ghaeilge, 1989
Tionscadal Céime, Ollscoil Luimnigh

Córas Fuaimeanna Na Gaeilge
Micheál Ó Siadhail & Arndt Wigger
Institiúid Ard-Léinn, BÁC

Graiméar Gaeilge Na mBráithre Críostaí, 1960
M.H. Mac An Ghoill & A Mhac,Tta
Sráid Uí Chonaill Uachtarach, BÁC 1

Gramadach Na Gaeilge agus Litriú Na Gaeilge, 1968
An Caighdeán Oifigiúil
Oifig An tSoláthair, BÁC

Nuachúrsa Gaeilge na mBráithre Críostaí, 1970
M.H. Mac An Ghoill & A Mhac,Tta
Sráid Uí Chonaill Uachtarach, BÁC 1

Réchúrsa Gramadaí, An 3ú hEagrán, 1963
Brian Mac Giolla Phádraig
Longman, Brún agus Ó Nualláin.

Correcting Syntactic Errors in German Sentences: A Chart-Based Approach

Ian M O'Neill[1]

Michael F McTear[2]

[1]BIS Beecom International Ltd

[2]University of Ulster

Abstract

This paper describes the design and implementation of a natural language processing system for correcting syntactic errors in German sentences. The errors treated mainly involve word order, but errors of omission are also covered. Chart-based parsing was used as a technique for detecting errors and suggesting possible corrections. The system was implemented in Prolog. In the paper the basic techniques of chart parsing are outlined, followed by a discussion of how these techniques can be usefully employed in association with correction strategies to detect and correct ill-formed sentences. The paper concludes with an evaluation of the present system and some indications for further research.

1. Introduction

This paper outlines some of the issues confronted and the solutions chosen in the course of a project to design and implement a natural language processing system for correcting syntactic errors in German sentences. Ultimately such a system would be used in situations where a non-native speaker of German might be expected to falter over problems of German word order. The prototype system which is described here was implemented in Prolog and uses chart-based parsing (in combination with heuristics for typical error patterns in German) as an efficient means of correcting sentences.

The paper is structured as follows. In the next section we will outline the main problems which arise in German syntax for non-native speakers (particularly those whose mother tongue is English). This will be followed by an overview of chart-based parsing. The main part of the paper will present the correction strategies which were applied to German sentences. The paper concludes with an evaluation of the present system and an indication of how it might be further developed.

2. The German Problem

While German bears close similarity to English in many respects, word order is one area in which the two languages diverge markedly. German demands, for example, that in a main clause which is a statement the verb should always be the 'second idea', though not necessarily the second word. Elsewhere, German has a tendency to move verbs or parts of verbs to the end of clauses. For example, in that same declarative main clause, if the finite verb happens to be an auxiliary that takes a past participle, or if it is a modal auxiliary that takes an infinitive, then, generally speaking, the past participle or the infinitive will be pushed to the end of the clause. Matters are complicated further if a modal construction is to be used in the perfect tense in a main clause, for then two infinitives will be found at the end of the clause, and in the reverse order to that which the English speaker might expect. In a subordinate clause, the positioning of verbal elements becomes yet more un-English, for then the finite verb itself goes to the end of the clause, immediately preceded by any infinitive or past participle dependent upon it - unless, that is, there are two or more dependent infinitives, in which case the auxiliary will precede them! Figure 1 below gives some examples of German sentences, with their English equivalents, which illustrate these differences in word order.

1. The finite verb as 'second idea' in a main clause

Heute geht er in die Stadt.

Today goes he to town

2. Past participle at the end of a main clause

Er ist in die Stadt gegangen.

He has to town gone

3. Infinitive following a modal verb in a main clause

Er will in die Stadt gehen.

He wants to town to go

4. Modal and past participle in a main clause

Er hat in die Stadt gehen wollen.

Er has to town to go want(ed)

5. Subordinate clause - finite verb at end

Das ist der Mann, der in die Stadt geht.

That is the man who to town goes

6. Subordinate clause - finite verb at end preceded by past participle

Das ist der Mann, der in die Stadt gegangen ist.

That is the man who to town gone has

7. Subordinate clause - an auxiliary and two infinitives

Das ist der Mann, der hat in die Stadt gehen wollen.

That is the man who has to town to go want(ed)

Figure 1. Word order in German main and subordinate clauses

This paper will deal largely with the question of repositioning misplaced verbal components of a clause. However, some thought will also be given to the matter of incorrectly ordered German adverbs, which in general must follow the sequence 'time, manner, place' if they occur after the verb in a main clause, or before the verb in a subordinate clause. Likewise, some consideration will be given to the problem of missing words: for example, while a relative pronoun ('that', 'which') will often be omitted in an English construction, this can never be the case in German; similarly certain set expressions will demand the use of the definite article in one language whereas it is uniformly omitted in the other.

3. Choosing a parsing strategy

There are two fundamental issues in developing a system for analysing sentences in natural language:

1) choosing a formalism (grammar) to represent the valid syntactic structures of the language;
2) choosing a parsing strategy which takes this grammar and a set of sentences as input and which then outputs an analysis of the sentences in terms of their syntactic structures.

In the case of the present system, a further issue (to be described in greater detail later) was:

3) choosing a method for recognizing ill-formed input and outputting a corrected version of the input.

Although many different formalisms are now available to computational linguists for representing syntactic structure (see [1] for a review), in this project a simple formalism - context-free phrase structure grammar (CF-PSG) was selected, the reason being that the main aim of the project was to focus primarily on the parsing and correction strategies.

Chart-based parsing was selected as a technique for analysing the structure of sentences using these grammar rules (for a useful overview of parsing strategies, see [2]). One of the main advantages of chart-based parsing is that it is possible to keep a track of intermediate results [3,4,5]. A parser working top-down and backtracking as appropriate according to the normal Prolog control structure is likely to be inefficient as it fails to keep intermediate results. So, for example, if a grammar rule requires a noun phrase followed by a verb phrase and the parser succeeds in finding a noun phrase but is unable to find a suitable verb phrase, then the parser will backtrack and try another rule, perhaps having to look once again for a noun phrase. Thus the intermediate result - that there was indeed a well-formed noun phrase - is lost and the parser has to re-do work which had already been done. Chart-parsing overcomes this problem by keeping intermediate results - in particular, well - formed substrings that may at a later point in the parsing process contribute to the analysis of the sentence.

4. Chart-parsing: an overview

Grammatical structures can be represented by directed acyclic graphs (DAGs), the most commonly used form being the tree. It is possible, using different tree diagrams, to show alternative structural analyses of a sentence, as in Figure 2. The tree to the left represents a structure in which 'they read the book on horses' may be taken to mean that 'they read the book, which was about horses', while the tree to the right represents a structure meaning 'they read the book while on horseback'.

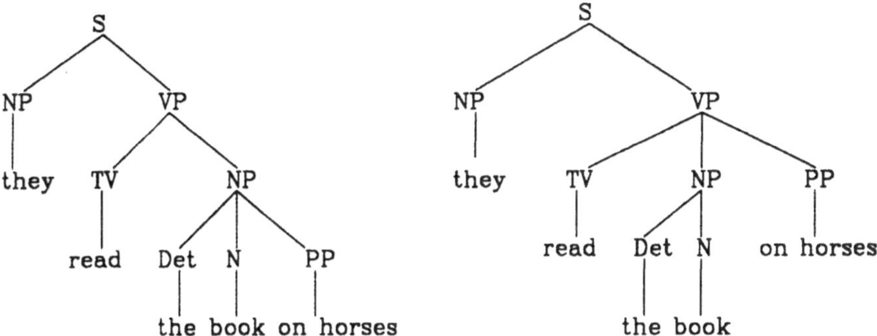

Figure 2 Alternative interpretations of 'they read the book on horses'

It is more difficult when we wish to represent alternative sub-structures using trees and for this reason the chart is a more useful form of representation. It is simple to convert a tree-like representation of grammatical structures into a chart-like one. In a tree the nodes carry the labels which identify the categories of structures or of individual words. In a chart vertices are placed rather like fence posts between the words of the string to be analysed, so that each word has a vertex on either side. The category of the word is indicated on the arc (or edge) which when drawn from vertex to vertex spans that word. Similarly, longer edges may span a number of shorter edges, so that the longer edge comes to represent a more complex structure comprising (at least in part) the grammatical categories represented by the edges which it subsumes. The advantage of this form of representation is that an appropriate edge can be added to the chart as soon as the components of a particular category have been identified, regardless of whether or not such a category may eventually form part of a more complex structure. Likewise alternative interpretations of a word, or of a conjunction of categories, may be indicated by drawing several edges between the same pair of vertices. Figure 3 represents the information contained in Figure 2 as a well-formed substring table (WFST).

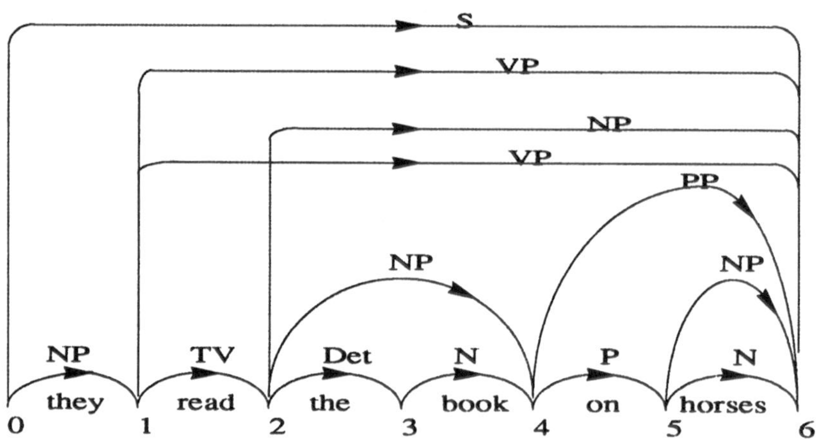

Figure 3 A well-formed substring table for 'they read the book on horses'

Converting a WFST to an active chart is largely a matter of changing the way in which the edges are labelled. The edges of a WFST indicate only those structures and individual components which are actually present in the input string. For a parser which will be adding edges one at a time and checking to see if a category (of phrase or word) represented by an edge can be subsumed into a more complex category, it would be advantageous to have a form of labelling which indicates whether a category is complete, or still requiring components, or indeed whether the presence of a category is purely conjectural, though likely in particular circumstances. Figure 4 presents a fragment of a chart which contains this sort of information.

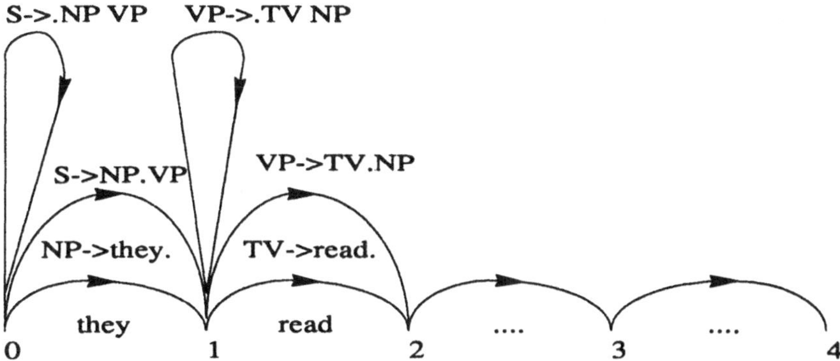

Figure 4 Active and inactive edges in a chart fragment

The manner in which the edges of the chart are labelled is worthy of particular attention, for, as has just been suggested, the presence of an edge may indicate either that a syntactic structure is complete, or that it is partially complete: moreover, an edge may also signify that the presence of a particular structure is possible though as yet entirely unproven - and as will presently be seen, this speculative quality of the chart is of great significance to the formulation of parsing algorithms. It is largely the labelling convention which facilitates this representation of complete, incomplete and conjectured structures. The following exemplifies the 'dot notation' used to label the edges of the active chart:

(i) S -> .NP VP

(ii) S -> NP.VP

(iii) S -> NP VP.

The notation is easily interpreted: any component which follows the dot is yet to be found, while any component which precedes the dot has been found already. In addition, the form of the notation makes it clear precisely which components are required to form a more complex structure - thus, in the above examples, it is obvious that a noun phrase followed by a verb phrase is one combination of components which may be deemed to constitute a sentence. Furthermore, although the active chart is essentially an acyclic directed graph, rules of acyclicity are relaxed to accommodate the purely speculative edges, which are allowed to cycle back to the vertex at which they originate - there being little point in depicting a purely speculative edge as spanning a number of vertices. Edges which represent an incomplete syntactic structure - edges which, in other words, would be labelled with one or more elements following the dot in the 'dot notation' - are known as active edges: as one might expect, edges which represent complete syntactic structures are referred to as inactive edges.

5. Using chart parsing to detect ill-formed input

If inaccuracies and omissions in syntactic structures are to be detected, and useful suggestions made as to how such ill-formedness might be corrected, then evidently some means of determining accurate syntax, and predicting how accurate syntax might be established, is essential. Already, the introduction of the concept of active and inactive edges has provided a basis for recording syntactic completeness and for determining which sentence components could (once they were available) be usefully combined with already identified elements. First, though, one must consider a method of analysing a given input string so as to build more complex syntactic structures from the individual lexical items available. This can be accomplished by following two relatively simple parsing rules - the 'fundamental rule' and the 'top-down rule'.

5.1 The Fundamental Rule

The fundamental rule states that if an active edge requiring a particular component meets an inactive edge which represents that component, then a new edge, drawn from the leading vertex of the active edge to the closing vertex of the inactive edge, may be added to the chart. The fundamental rule is thus a mechanism which allows smaller syntactic structures to be encompassed within larger ones, so that eventually all the individual components of an input string will (if it is at all possible) be drawn into a single sentence. However, for the fundamental rule to function it requires a suitable mixture of active and inactive edges. The second of the two parsing rules will indicate the circumstances in which active edges may profitably be added to the chart.

5.2 The Top-Down Rule

It is a relatively simple matter to label with inactive edges the individual words of the input string (and, conveniently, the chart allows the same pair of vertices to be spanned by as many labelled edges as are necessary to convey the various syntactic interpretations of a word or phrase). However, as should now be evident, for the fundamental rule to be applied (and for the parse in consequence to progress) the chart must contain active edges which require at least some of the components represented by the inactive edges. The top-down rule dictates firstly that a purely speculative active edge (or indeed a number of such active edges) should be allowed to cycle out from and back to the opening vertex of the chart and that this edge (or these edges) should represent any syntactic category capable of spanning the entire chart. In most cases this means that the first active edge to be added will speculate as to the existence of a sentence, perhaps taking the form

S -> .NP VP

which in effect conjectures that the newly input string may contain sufficient suitable elements to build a noun phrase, followed by a verb phrase, which would then constitute the chart-spanning sentence.

The process of adding these speculative active edges for chart-spanning structures is performed only once. Thereafter the top-down rule demands that for every active edge added to the chart, a further, again purely speculative, active edge should be added for the first element sought. Thus, in the above example, where a noun phrase is required, another edge, labelled

NP -> .Det N

might cycle out from and back to the vertex on which the initial active edge ended - in other words both active edges cited above will begin and end on the opening vertex. The proliferation of active edges which is caused by the top-down rule brings with it the major advantage that, as the fundamental rule finally comes into play, the inactive edges which represent the individual words of the input string are drawn 'purposefully' into more complex semantic structures whose place in an all-embracing sentence has already been envisaged: the 'layers' of active edges, from the mooted sentence downwards, constitute one or more syntactically accurate sentence structures into which the individual lexical items may be slotted as appropriate. Despite the fact that such frameworks may be devised even in the absence of the individual component words (such is the disadvantage of working from the top down!) the predictive quality of such an approach proves invaluable in a sentence checking system whose aim is either to suggest how a syntactic structure may be completed, or how those items which are present may be rearranged to create well-formed output from ill-formed input.

6. Correction strategies

6.1 Overcoming parse failures, and finding the missing word

A useful step towards devising a correction strategy is to examine the state of a chart (on which the top-down and fundamental rules have been brought to bear) when the parse eventually fails. If an input string is parsed from left to right, then a parse failure due solely to the absence of a word would mean that the last active edges to be added to the chart (those furthest to the right of the chart) would contain in their number one edge that would indicate the missing category: the chart would not be 'extended' (no other edges would be added) beyond these active edges since a suitable inactive edge, which could be exploited by the fundamental rule, would be absent. An obvious correction strategy resulting from this observation is firstly to create inactive edges corresponding to those required by the last active edges, secondly to make a space in the chart, if necessary, so that the newly created inactive edges can be inserted (one at a time), and thirdly to ascertain which of these new inactive edges best furthers the parse (to the extent of allowing a complete sentence to be parsed).

If the resultant parser cum word finder is fed sentences in which a word is obviously missing, then, provided that the intended syntactic structure is catered for in the CF-PSG associated with the parser, it should be a straightforward matter for the system to determine which syntactic element is missing (i.e. which category of word, if inserted at a particular point in the input string, would allow the entire string to be parsed as a well-formed syntactic entity.) As will be discussed presently, although such a word-finding strategy is virtually assured of success in the contrived circumstances just described, it is not an especially discriminating mechanism - so that complications would obviously arise if a word had simply been misplaced rather than actually omitted, or if it were possible by turns to create a number of syntactically correct but semantically quite different sentences while still inserting only one category of word at a time into the input string.

6.2 Finding the right word order

In the context just outlined, a simple examination of the active edges left unexploited at the point when the parse failed was an adequate basis on which to determine which element, if added to the input string, would produce a syntactically correct sentence. However, such a straightforward, 'mechanistic' solution is less readily attained in the case of words being misplaced rather than simply omitted. Now one must take into account not only which syntactic category or categories are required to continue a parse, but also which categories are actually available in the input string. In order to avoid a random rearranging of sentence components, in the hope that one syntactically sound sentence might emerge, it is helpful to predict the kind of word order errors that might occur, and to facilitate their correction by means of some sort of check list - this indeed is the approach adopted for Exeter University's natural language processor, LINGER, which performs a series of ad-hoc 'checks' en masse on any syntactically defective sentence [6].

In the case of German being spoken by an English native speaker, the word order problems as listed in Figure 5 are likely to occur. On the basis of these observations a number of 'mal rules', or rules of thumb, may be formulated, each rule anticipating one of the word order problems just described, and recommending a course of action to be taken - in other words the correction strategy to be followed. It may be that a number of mal rules will call on what is essentially the same correction strategy, the strategy varying only in the number of sentence components it is expected to handle: thus, for example, the mal rules relating to problems 1,2,3,6,7,8 and 9 of Figure 5 will all draw on a variation of the correction strategy which moves components to the end of a clause. To clarify the point further, it is worth considering the way in which two possibly troublesome components are identified and moved.

In a main clause:

1. A misplaced past participle.
2. A misplaced infinitive following a modal verb.
3. Two misplaced and possibly incorrectly ordered infinitives following an auxiliary (the perfect tense in a modal construction).
4. The finite verb being other than 'second idea'.
5. Two or more adverbs incorrectly ordered.

In a subordinate clause:

6. A misplaced simple finite verb.
7. An auxiliary and a past participle, misplaced and possibly incorrectly ordered.
8. A modal auxiliary and an infinitive, as number 7.
9. An auxiliary and two infinitives, as number 7.

('Misplaced' means that an item is incorrectly positioned in relation to the rest of the clause, and 'incorrectly ordered' indicates that two or more items are incorrectly positioned in relation to each other).

<u>Figure 5. Typical word order errors in German</u>

In broad terms, if a mal rule indicates a potential problem, such as an auxiliary and a past participle not being correctly positioned and correctly ordered at the end of a relative clause, then, in any ill-formed string in which these components can be identified, they will be removed, and the string examined again to see where they might usefully be reinserted. In effect, one is assuming the worst: English speakers often have difficulty with the auxiliary-past participle construction in a relative clause, so perhaps this problem has arisen in the ill-formed string under consideration. Of course, if the potentially troublesome construction has in fact been used correctly, then its rearrangement and/or reinsertion will serve no purpose (indeed, it is likely that the removed components will be returned to their former positions). In these circumstances a parse will fail even after the application of a particular mal rule, forcing another mal rule to be considered.

Let us examine this correction strategy more closely, specifically with regard to the mechanics of the active chart and the related correction algorithm.

Firstly, if a component is to be removed (temporarily) from an input string, then the inactive edge which represents it in the chart must be removed, and the chart closed around the ensuing gap. (In the case of two components being removed, the algorithm will repeat this sequence.) The chart can then be parsed afresh, and is examined for any active edge which indicates that all of the components just removed are required for the completion of a stipulated structure (if an auxiliary and a past participle have been removed, then the mal rule may demand that they form part of a relative clause when reinserted). All such active edges will be gathered into a 'candidate list' and sorted, so that the active edge with the highest numbered final vertex heads the list and is thus selected for further processing: it is the aim of this particular correction strategy to move sentence components to the end of a clause, and so it is important that, from among the candidate active edges, one is chosen which will allow the newly removed components to be usefully replaced as far to the right of the input string as is possible. (The limitations of this strategy will be discussed presently.) New inactive edges representing the components to be reinserted into the chart are now created on the basis of the information contained in the chosen active edge, the chart is again rearranged - now to accommodate the newly formed inactive edges - and the complete chart is reparsed. An examination is made to see if the parse has produced an all-embracing sentence, and if so the user can be informed of the nature of the original syntactic error, and how its correction can be achieved.

The majority of the problems listed in Figure 5 are resolved by the above correction strategy and its variants. However, among the problems still to be tackled is that of the verb being other than 'second idea' in the input string. In German, several different types of constituent - such as an adverbial word or phrase, an object noun phrase, or a subordinate clause - can be placed as first idea and then must be followed by the main verb as second idea.In this project we were concerned only with adverbials or subjects as first idea. In many cases, if the verb is not already second idea in a simple declarative sentence, then the most obvious correction strategy is to move that verb. Such an operation is readily performed on the basis of a correction strategy not unlike that just described - a notable point being, however, that an algorithm to move a verb to the position of 'second idea' does not need a list of candidate active edges: provided that the grammar outlines sentences whose 'first idea' is a clearly identifiable expression of time, manner or place, e.g. an adverbial phrase, then it will be a simple matter to move the verb to the position immediately following the opening phrase.

However, in certain circumstances - for example if the subject is followed immediately by a relative clause - then to preserve clarity of meaning it may prove advisable to move the opening adverbial element rather than the verb if the latter is not 'second idea': thus *Heute Gerd, der laut schnarcht, arbeitet in der Stadt* would become *Gerd, der laut scharcht, arbeitet heute in der Stadt* (the verb of the main clause being *arbeitet*, the expression of time *heute*).

So as to move the adverbial element only when this is strictly necessary (such a movement amounts, after all, to a stylistic alteration) it is possible first to remove the opening adverb, then to reparse the remains of the sentence (which, in the case of a sentence similar to that just presented, will constitute an accurate sentence in its own right: *Gerd, der laut schnarcht, arbeitet in der Stadt*) so as to ensure that the input string does indeed contain a relative clause after the opening noun phrase. If this is the case, then the correction algorithm finds an active edge that indicates a place for the removed item to be reinserted, checks (on the basis of vertex numbers) that this active edge does not lie within a relative clause (since the aim is to remove an awkward element from the start of a main clause to the body of that same clause) and only then performs the reinsertion. In the current implementation this particular correction strategy would be tried before that in which the verb itself is moved - such sequencing ensures that the eventuality of there being a troublesome relative clause is taken into account before the simpler solution is chosen.

The question of how to order the constituent elements of adverbial groups is fairly easily resolved. If it is accepted that, as a general rule, adverbs occurring together will follow the sequence 'time, manner, place', then the task of reordering will simply be a matter of swapping one inactive edge for another - again assuming that the grammar defines each adverb as falling into one of these three categories. In an actual implementation, however, a number of steps will need to be followed before swapping commences. Firstly, as this algorithm is intended to rearrange adjacent pairs or unbroken triplets of adverbs then checks will need to be made that the adverbs found are indeed contiguous, and this amounts to a simple procedure of examining the vertices of the relevant inactive edges to see (in the case of a pair) that one vertex is common to both inactive edges, or (in the case of a triplet) that the leading vertices of the three inactive edges can be arranged into a continuous sequence of integers. The check for a triplet should be performed before the check for a pair, so that a complete conjunction of adverbs will be identified before any correction routine is chosen.

7. Assessment

For a parsing system, such as the one outlined here, to be of genuine value in an educational context, then clearly it would not only have to cope with a very broad range of ill-formed input (and recognise the intended well-formed constructions!) but in addition would have to allow input, and present its recommendations, in a very user-friendly form. Unfortunately, the system which emerges from the rather informal, high-level specifications outlined here is some distance from achieving either of these objectives. For example, an interface which relies heavily on the terminology used in the context-free phrase structure grammar, and on the system of vertices by which the chart parser keeps track of the position of individual words and phrases, will give rise to a very rudimentary interface indeed - advising, for instance, that "the <ivprt> of the main clause (word number 6) should be moved to the end of the main clause". Clearly it would take an already competent grammarian to interpret this message as an invitation to move the past participle of an intransitive verb to the end of a sentence.

As has been suggested, the correction strategies themselves are far from complete. The word finding mechanism was formulated on the assumption that a word was indeed missing, and not for example, that the incorrect form of a word had been included or indeed a word misplaced: a means of enhancing the word finder's abilities and combining them with the efforts of the more comprehensive (sentence re-ordering) 'sentence checker' would thus be of the utmost importance. The sentence checker itself falls far short of the full flexibility of spoken and written German, largely on account of its rather limited CF-PSG: in the brief time-span available the project necessarily simplified much of the flexibility that German enjoys on account of its clear morphological representation of grammar and case - this allows, for example, a sentence to begin with direct and indirect objects (never mind infinitives or past participles) while maintaining the main verb in its position of 'second idea'. Likewise no consideration was given to more colloquial German usage (which for example allows adjuncts of time, manner, or place to be added after the main verb of a relative clause) or to the fact that a relative pronoun may for instance refer to a remote antecedent contained within an already completed relative clause, or perfect/pluperfect construction. Similarly, the fact that input was limited to, at most, one main and one relative clause, meant that moving lexical items to the end of clauses remained an easily manageable task, governed principally by highest numbered vertices. These and other limitations naturally inhibit the authenticity of the input strings. Certainly a reworking of the grammar and lexicon to allow the use of features would add a greater degree of processing efficiency and allow details of number, gender, case and semantic area to be incorporated into more sensitive correction strategies.

8. Conclusion

Despite its inevitable shortcomings and generalisations, this project to design and implement German sentence checking mechanisms provided many useful insights into the working of the active chart, and ways in which such a chart can be developed, analysed and manipulated. Furthermore, the project revealed the extent to which syntactic analysis alone could be used to convert ill-formed to well-formed input. If this work stimulates others to investigate further the partially resolved or as yet fully unresolved issues which have just been discussed, then it may indeed be deemed to have been well worthwhile.

References

1. Sells P. Lectures on Contemporary Syntactic Theories . CSLI Lecture Notes No 3, Chicago University Press Chicago, 1987
2. Shapiro S., Eckroth,D. (eds.) Encyclopedia of Artificial Intelligence. Wiley New York 1987
3. Allen J. Natural Language Understanding. Benjamin/Cummings Menlo Park California 1987
4. Gazdar G., Mellish C. Natural Language Processing in Prolog. Addison-Wesley Wokingham 1989
5. Winograd T. Language as a Cognitive Process, Volume 1. Addison-Wesley Reading Massachusetts 1983
6. Yazdani M. An English Tutor Project Report (1987 - 1989) . Employment Department Group Training Agency Sheffield 1989

TUIG - towards a parser for Irish

Dáire Mag Cuill
Dublin College of Catering
Cathal Brugha Street
Dublin 1

Kevin Ryan
University of Limerick
Limerick

Abstract

TUIG is a natural language processing system for Irish which creates a representation of the information contained in declarative sentences in Irish using an approach combining Phrase Structure Grammars and Case Frame Grammars. The syntax recognised by TUIG is based on research carried out during the 1960s into the occurrences of sentence structures in spoken Irish.

1 Introduction

TUIG is a natural language processing system for Irish which creates a representation of the information contained in declarative sentences in Irish using an approach combining Phrase Structure Grammars and Case Frame Grammars. The domain chosen for TUIG was weather reports.

The version of TUIG described in this paper has two phases - a pre-parsing phase and a parsing phase. A less developed semantic processing phase is beyond the scope of this paper, as are TUIG's mechanisms for "learning" new words to add to its database.

The first phase, pre-parsing, ensures that all words are rendered in their basic form as found in the lexicon and takes note of any deviations where they might be relevant to understanding. The second phase, parsing, identifies the phrases which make up the sentences and matches the sentence structure to one of over one hundred potential structures.

This paper first explains how the structure of the parsed sentence is represented, and then describes in detail the various stages of TUIG's processing. This is followed by examples of TUIG's work, and a number of conclusions drawn from the work.

TUIG was implemented in Edinburgh Prolog and runs on an Apple Macintosh.

2 Representing the Parse Structure

The parse structure derived is a linear representation of a parse tree, where nested brackets show the relationships. For example the sentence:

déanann sí é = does/makes she it
she does/makes it

gives

'S'('+A'('B'(((déan,1),'A'('PRO'(f3)))),'A'('PRO'(m3))))

In this example, the brackets preceded by the S encloses the sentence; +A means that the basic sentence structure is followed by a noun phrase (A = *ainmfhocal*); next is the verb phrase (B = *briathar*) where "1" stands for the present tense (*aimsir láithreach*), which is followed by its subject verb phrase (A again); and finally comes the noun phrase referred to by +A above.

This representation has a number of advantages: the sentence structure (see below) - in this case B+A - can be seen easily at the beginning; and the nesting of brackets has the same representational power as a standard parse tree while being linear and easier to manipulate. Indeed programs exist which produce trees from such parenthesized input.

3 The stages of pre-parsing

3.1 Tokenisation

This processing of the input into correct tokens is the first stage of TUIG's pre-parsing. The changes to be made fall into four groups: tokens to be merged; tokens to be split; tokens to be changed; and tokens to be discarded. Some changes are actually the product of more than one of these types.

The tokens which must be merged are those which logically pertain to one unit of the sentence, typically only one would be a lexigraphic entry in its own right. For example the Irish text *an t-oileán* would be tokenised by Edinburgh PROLOG into four tokens: "an", "t", "-", and "oileán". Only "an" and "oileán" are actually words, but "t" pertains to "oileán", and actually contains the information that this noun is in the nominative case. On the other hand "-" is only a typographical convention which shows that the word is not "toileán". In any case this information must be stored along with the token it pertains to, and not as separate tokens.

The sample list of these changes is given below along with a selection of the details of the semantic information which can deduced even at this stage.

The tokens which must be split to produce more than one token are those pertaining to words which by convention are run together in written Irish. For instance orthographical convention states that *atá* is one word whereas *a bhfuil* is two, despite the fact that they are respectively the dependent and independent relative forms of the verb *bí* (present tense) prefixed by the relative particle *a*. In order to process the input sensibly the treatment of the particle must be uniform.

The third type of change is where one token is substituted for another. The basic case of this is where elision is eliminated and the original complete word restored. For example *ba* is often elided to *b'* and this is changed back. In the example of *a'* , this is replaced by "à" because it is not always a case of elision and further processing must be done to see if it represents elided *an* or a verbal particle.

To avoid confusion or ambiguity, since commas are used by PROLOG, commas are replaced at this stage by "x_camóg", a token which could never occur otherwise.

The final type of change is where tokens are discarded altogether. For example hyphens other than those dealt with by previous rules are discarded because they usually occur in situations such as unnecessary separators between *an* or *ró* and adjectives.

Some examples of this tokenisation are given below:

Text	Example	Default Tokens	Correct Tokens	Occurrences
d'<...>	*d'fhág*	d ' fhág	d_fhág	• "do" is prefixed to past tense or conditional verbs - elision occurs before vowels or "fh"
h-<...>	*hoileáin*	hoileáin	h_oileáin	• "h" is prefixed to plural nominative or accusative case nouns beginning with a vowel after the article "na"
	hoileán	hoileán	h_oileán	• "h" is prefixed to nouns beginning with a vowel after "a" - the third person singular feminine possessive particle
atá	*atá*	atá	a tá	
sa	*sa*	sa	i an	• "sa" is a contraction of "i" + "an"

Since the circumstances described above which give rise to token changes are disjoint in no case can more that one tokenisation change be relevant, and there is therefore no need to be concerned with the problem of which rule to choose. Likewise once a replacement has been made it is possible to move on to the next token in the default list and to continue in this way until the list has been fully completed. If no change is needed then the default token is passed on as it is.

3.2 Tagging

The tokens produced after the tokenisation stage are not all words that will be recognised by the parser. In some cases, such as those new tokens created, the tokens contain extra characters - for example "d_". In other cases, where the words have either *séimhiú* or *urú* as a mutation, they will not have the same form as that stored in the database.

These initial mutations - *séimhiú* and *urú* - which are typical of Celtic languages and involve a modification of the initial consonant sound, can occur for a large variety of reasons, some of which are explained in this paper. More details are available in grammatical references [ÓS89] [MGPh63].

Tagging takes account of these features, and replaces each token with an ordered pair - where the first element is a word and the second a single character tag. This tag contains the information known about these words at this stage. Words which have no specific tag are given "b" as a default tag for consistency - so that all words have a tag.

Tokens which have *séimhiú* s have the token "s", and tokens with *urú* have "u". In the case of those features identified in the tokenisation stage the tag is that character preceding the underscore, for example the token "d_fhág" gives the ordered pair (fág,d). This yields the set of tags {d, n, t, h, ú, x} - some of which are explained above.

The *séimhiú* s and *urú* s are identified, removed and tagged as appropriate by a set of rules. The characters are specified in ASCII. Note that words beginning with "bhf" can be considered either as having a *séimhiú* or an *urú* but will be considered an *urú* for our purposes.

4 Parsing Sentence Components

Buntús Gaeilge [ÓH66] gave a list of nineteen sentence components for Irish which were then used to specify the permitted sentence structures which we present in the next section. Some of these sentence structures require a specific instance of a component, and others a general one.

Based on these components, and adapting some of them to make them more suitable to the task in hand and to automation, the following table lists the sentence components which were determined.

A - Noun Phrase	B - Verb Phrase
C - the copula *is*	D - Adverbial Phrase
E - the pronoun *ea*	F - pronoun
H - Verbal Noun	J - adjective or adjectival clause
K - Verbal Adjective	P - a form of *ag* showing ownership
Q* - Query Word	R - Preposition or Prepositional Pronoun
S - Special uses of *seo / sin*	U* - Interjection or exclamation
V* - Vocative noun	W - conjunction with following dependent verb
X* - Non-Standard relative or other clause	
Y - Independent uses of the verb *bí*	Z - conjunction with following independent verb

* These are not implemented in the current version of TUIG.

These are further modified by addition of digits or letters in some of the constructions in order to describe the construction more fully - and allow for those sentence structures where specific instances of particular components are required. The following sections detail a selection of these classes of sentence constituents, including their subsets, and describes of how each is dealt with in TUIG.

A Noun phrase

The noun phrase may be a noun expression, a single noun or even a pronoun, although these are dealt with separately as E and F in Buntús Gaeilge [ÓH66]. The noun phrase is taken along with any adjectives, which are detailed under J. Noun phrases are subdivided as follows:

A	General	includes "ordinary" Noun Phrases and all of the following
A1	Clásal (Clause)	A whole clause can be the Noun Phrase
A2	Number	
A3.1	*fios*	A3.* are nouns occurring in specific
A3.2	*sin*	contexts and are treated as
A3.3	*féidir*	special_nouns
A3.*	etc.	
A4.1	Pronoun	
A4.2	Pronoun+ *sin* | *féin* | *seo*	(these demonstrative adjectives are treated as adjectives under J)
A5	Proper Noun	

Based on these breakdowns TUIG allows the following recursive structures:

Noun Phrase ---> Noun Phrase + Conjunction + Noun Phrase
Noun Phrase ---> Determiner + Noun Expression + Genitive Noun Expression
Noun Phrase ---> Determiner + Noun Expression
Noun Phrase ---> Noun Expression + Determiner + Genitive Noun Expression
Noun Phrase ---> Noun Expression

Noun Expression ---> Noun Expression + Adjective Expression
Noun Expression ---> Noun (Nominative Singular)
Noun Expression ---> Noun (Nominative Plural)
Noun Expression ---> Preceding Adjective + Noun Expression
Noun Expression ---> Pronoun

Genitive Noun Expression ---> Genitive Noun Expression
+ Genitive Adjective Expression
Genitive Noun Expression ---> Noun (Genitive Singular)
Genitive Noun Expression ---> Noun (Genitive Plural)
Genitive Noun Expression ---> Preceding Adjective + Genitive Noun Expression

Noun ---> Existing Noun
Noun ---> Special Noun (as in A3.* above)
Noun ---> New Noun

Nouns are defined in the database in four forms: singular and plural, nominative and genitive. Other cases are based on the nominative. New Nouns are those added to the database since the beginning of the session. Special Nouns have only one form which is always used. In fact they very rarely occur in anything other than the nominative singular in petrified phrases, but if necessary they could be entered fully as other nouns in the database.

A separate particle defines h_nouns as those in the database beginning with h - this is used in the tokenisation process to establish whether a word beginning with a "h" is one of these few nouns which actually begin with the letter "h", or whether it is a "h" added to the beginning of a word beginning with an initial vowel affected by the immediately preceding context.

Possible ungrammaticalities which would be allowed by the grammar above, such as a Noun Phrase being a Determiner followed by a Pronoun, are checked for explicitly. A list of pronouns is also included.

B Verb Phrase

The Verb Phrase usually includes the subject, although in the case of a *saorbhriathar*, or autonomous verb which corresponds to the passive voice, there is no explicit subject. Also in some cases the verb and pronoun subject are one word.

B General
B1 Verb + Noun Phrase
B2 Verb which includes a pronoun
B3 Autonomous Verb

This is implemented in TUIG as:

Verb Phrase ---> Autonomous Verb Expression
Verb Phrase ---> Verb Expression + Noun Phrase (subject)
Verb Phrase ---> Verb (which includes a pronoun as subject)

Verb Expression ---> Negative Verb
Verb Expression ---> Verb

Autonomous Verb Expression ---> Negative Autonomous Verb
Autonomous Verb Expression ---> Autonomous Verb

Verb ---> Existing Verb
Verb ---> New Verb

Autonomous Verb ---> Existing Autonomous Verb
Autonomous Verb ---> New Autonomous Verb

Each instance of a verb is stored in the database giving the verb as it occurs, the root of the verb, its tense and its tag. The tag is given as "_" which matches everything since is it is of no importance, but in some cases only a *séimhiú* or *urú* distinguishes two forms of the verb. New Verbs are those added to the database since the beginning of the session.

Autonomous Verbs are treated in the same way.

D Adverbial word, phrase or clause

This classification includes not only standard adverbs, but also any other verb qualifier.

D General
D1 A whole clause can be the Adverbial Phrase
D2 Compound adverb ie. a list of adverbs
D3 Noun as adverb
D4 Adverbs composed from *go* + <adjective>

This is implemented in TUIG as:

Qualifier	--->	Qualifier + Qualifier
Qualifier	--->	Other Qualifier
Qualifier	--->	Adverb + Qualifier
Qualifier	--->	Adverb
Adverb	--->	*go* + Adjective (one or more)

A list of Other Qualifiers which are not adverbs formed from adjectives is implemented. Adverbials of time would be an example of this type.

H Verbal nouns

Verbal Nouns, such as *déanamh*, are treated separately from verbs and are most often prefixed with specific prepositions, such as *ag*, which have different shades of meaning. These prepositions are described under R.

Verbal Nouns are stored in the database with the the root of the verb and the tag. As was the case with verbs the tag can be stored as "_" since it has more than one possibility.

Verbal Noun	--->	Existing Verbal Noun
Verbal Noun	--->	New Verbal Noun

J Adjective(s) and adjectival clauses

The adjectives include Verbal Adjectives, separately listed under K [ÓH66] and similarly listed here for completeness. As well as those adjectives occurring as special, there is another unusual set. While the vast majority of adjectives follow the noun, there is a small subset of adjectives which always precede the noun.

J	General	
J1	Verbal Adjective	
J2	Preceding Clause	Other clauses acting as adjectives
J3	Following Clause	
J4.1	*fíor*	J4.* are adjectives occurring in
J4.2	*dóigh*	specific contexts and are treated as
J4.3	*maith*	special_adjectives
J4.4	*iomaí*	
J4.5	*cuma*	

In TUIG this becomes:

Adjectival Expression	--->	Adjective
Adjectival Expression	--->	Conjunction + Adjective
Adjectival Expression	--->	Emphatic Prefix + Adjective
Adjectival Expression	--->	Comparative Adjective

Genitive Adjectival Expression	--->	Genitive Adjective
Genitive Adjectival Expression	--->	Genitive Conjunction + Adjective
Genitive Adjectival Expression	--->	Emphatic Prefix + Genitive Adjective
Genitive Adjectival Expression	--->	Comparative Particle + Comparative Adjective

Adjective	--->	Existing Adjective
Adjective	--->	New Adjective
Adjective	--->	Verbal Adjective
Adjective	--->	Special Adjective
Adjective	--->	Demonstrative Adjective

Adjectives in general are defined as were nouns except they have five forms: as well as singular and plural, nominative and genitive, there is a comparative / superlative form. As in the case of Special Nouns, Special Adjectives and Demonstrative Adjectives have only one form which corresponds to all forms, although we should note that the comparative form would never arise.

The set of Adjectives which come before rather than after the noun is defined separately, and includes ordinal and cardinal numbers as well as adjectives such as *céad*, *príomh* and so on. The number types are differentiated by their tag - ordinal numbers will have the tag "ú", cardinals will have "b".

As with Special Adjectives, Verbal Adjectives are defined as having one form. Provision is also made for New Adjectives.

K Verbal adjectives

As already mentioned Verbal Adjectives are dealt with under J.

R Prepositional Phrase

A Prepositional Phrase can be either a Preposition followed by a Noun Phrase or a Prepositional Pronoun.

R General	
R1 *ann*	specific use of *ann*
R2 Preposition + Noun Phrase	
R3 Prepositional Pronoun	

R4	Prepositional Phrase + *sin* | *seo*	(recall that these are treated as demonstrative adjectives)
R5	*ach* + Prepositional Phrase	
R7.1	*ag*	R7.* are prepositions that
R7.2	*le*	precede verbal nouns
R7.*	etc.	

This is covered by:

Prepositional Phrase ---> Preposition + Noun Phrase
Prepositional Phrase ---> Prepositional Pronoun
Prepositional Phrase ---> Conjunction + Prepositional Phrase

The demonstrative adjectives, where they occur, are recognised within the definition of Noun Phrase. The specific Prepositional Pronoun *ann* = "in it / there", with its special meanings, is also recognised separately as are variants of *ag* .

The Prepositions which precede Verbal Nouns are specified separately - and are especially rich in semantic information. However full processing of prepositional phrases, and specifically the nuances often conveyed by *ann* and *ag* must be left to the development of a semantic processor.

Y Bí Verb Phrase

The verb *bí* occurs in many specific cases which differ from those where other verbs may occur. For this reason it must be treated differently. As in the case of other verbs a Bí Verb Phrase usually includes the subject although in the case of a *saorbhriathar*, or autonomous verb which corresponds to the passive voice, there is no explicit subject. Also in some cases the verb and pronoun subject are one word.

Y General
Y1 Verb + Noun Phrase
Y2 Verb which includes a pronoun
Y3 Autonomous Verb

This is implemented in TUIG as in the case of other verbs, but of course "new verb" is not relevant here.

Each possible instance of a the verb *bí* is already stored in the database giving the verb as it occurs, as well as its tense and its tag. The tag is given as "_" if it is of no importance, but in some cases only a *séimhiú* or *urú* distinguishes two forms of the same verb.

Autonomous Bí Verbs are treated in a similar fashion.

5 Sentence Structure

5.1 Background

For the research on which Buntús Gaeilge [ÓH66] was based analysis was made of 22,109 syntactic units - in loose terms sentences, though some would not meet the precise definition of a sentence because of having no verb or whatever. Among other things analysis was made of the structure of these units. The most common structure occurred 1,076 times; the next common 1,060 and so on.

Of these units 12,338 , or 60% of the total, were found to have structures that occurred at least 20 times, and these are detailed in the report. There were 133 of these structures. This cutoff point was defended since it covered all of the most common forms of speech. In fact lowering the threshold to 10 occurrences only added 8% to the proportion covered, and a full 15% of the structures recorded were represented by only one occurrence.

Almost all of the 133 constructs have been represented in TUIG, the exceptions being those involving interrogative words, exclamations, ejaculations and what are referred to in the report as "non-standard reflexive or other clauses". These accounted for 370 occurrences spread over 13 structures. They are not appropriate since TUIG only seeks to analyse declarative sentences.

It can fairly be assumed, therefore, that TUIG recognises at least 60% of declarative sentences in Irish. In fact the actual proportion recognised is higher as will be described later.

On closer study it became clear that many of these 120 remaining declarative structures were actually related, being variants of one another formed, for example, by adding a prepositional phrase or adverbial phrase. Due to the recursive nature of PROLOG it was possible to reduce these structures to four simple structures which could occur on their own, with 11 optional extensions which were appropriate to some sentence structures but not to others but could not occur independently. A simple structure with one or more extensions would then be termed a complex structure.

Examples of these are detailed below, along with a list of the structures identified by Ó Huallacháin which they correspond to. In the case where more than one extension could be used with a simple structure the compound structure is classified by the last extension in the sentence. For example in the case of **B**-A-R-D where A R and D are extensions of the simple structure **B**, the details will be found under -D.

5.2 Notation

Simple structures are written in bold type, **B**, and extensions in plain type, R. Where an extension is included in a complex structure it is prefixed with a hyphen, **B**-R, or in the case of extensions applied to the beginning of a sentence it is suffixed with a hyphen, D-**B**. Clauses are separated by forward slashes. The codes are those used already in defining the possible sentence components. The number of occurrences of each type out of the total of 11,968 sentences studied is also given. Below this list is given the TUIG structures to recognise this set of of sentences.

In some cases different occurrences of the same structure are given separately depending on which of the simple structures, as I have called them, they occur with. These have been taken together here and the number of occurrences has been given for each type.

Simple Structure	Occurrences	Example(s) and explanation
B	1076	Verb Phrase *Tuigim, Déanann sé* - This usually represents the verb and subject but in the case of autonomous verbs the noun is the object
Y	885	*Bí* Verb Phrase *Tá mé* - This may or may not include a subject, and again in the case of autonomous verbs the noun is the object
C	N/A	The copula *is* Although it does not occur independently it is the basis for a number of complex structures
CE	665	*Is ea* - A specific, frequent occurrence of **C**

For the purposes of TUIG **CE** is treated as a specific case under **C** rather than as a separate structure.

5.3 Extensions

Extensions fall into two groups: those extensions which may occur with any of the simple constructions and those extensions which may occur with only some of the simple constructions. This stage of TUIG allows them to occur more or less in any order, once each follows of precedes an appropriate simple structure. For economy of code all are defined recursively and any incorrect structures which "slip through" can be identified by a method detailed later. Some examples are given below.

5.3.1 Extensions which can occur with any structure

Extension -R Prepositional Phrase

	Occurrences	Example(s) and explanation
Y-R	335	*Tá hata - air* = There is a hat - on him
B-R	268	*Shuígh sé - ar an suíochán* = He sat - on the seat
B-A-R	229	*Chonaic mé - solas - ann* = I saw - a light - there
B-D-R	166	*D'iompaigh sé - thart - orm* = He turned - around - on me

Extension -D Adverbial Phrase

	Occurrences	Example(s) and explanation
B-D	343	*D'imigh sé - amach* = He went - out
B-A-D	163	*Thóg sé - é - go gasta* = He took - it - quickly
Y-D	152	*Tá - cinnte* = It is - surely
Y-R-D	117	*Tá blas searbh - orthu - go fóill* = There is a bitter taste - on them - yet

Extension -A Noun Phrase

	Occurrences	Example(s) and explanation
B-A	1060	*Déanann sé - é* = He does - it
B-D-A	182	*Crochfaidh sé - suas - an lampa* = He will hang - up - the light
Y-A-P-A	168	*Ní raibh - fhios - aige - an difear* (Could be written YPA) = There was not - the knowledge - at him (He did not know)
B-R-A	151	*Thug sé - dom - é* = He gave - to me - it
C-A-A	127	*B'fhéidir - é* = Perhaps - it (so)

Extension D- Adverbial Phrase

	Occurrences	Example(s) and explanation
D-B	30	*Ar ndóigh tiocfaidh sé* = Of course - he will come

5.3.2 Extensions which are specific to particular structures

As well as the extensions given above, some extensions can only properly occur with given basic structures or along with other extensions.

Extension	-J	Adjectival Phrase	Appropriate to **Y** and **C**
	Occurrences	Example(s) and explanation	
Y-J	280	*Tá an fear - mór* = The man is - big	
Extension	-P	*ag* showing ownership	Appropriate to **B** and **Y**
	Occurrences	Example(s) and explanation	
Y-A-P	233	*Tá - fhios - agam* = There is - knowledge - at me (I know) (This could be written YP and include the next example)	
Y-P	155	*Bhí siad - agam* = I had them	
Extension	-RH	Preposition followed by Verbal Noun	Appropriate to **B** and **Y**
	Occurrences	Example(s) and explanation	
-RH	256	*(Ní fhaca mé é) ag snámh* = (I did not see him) swimming	
B-A-RH	28	*Thriail sé - cloch - a chaitheamh* = He tried - a stone - to throw	
etc.			
Extension	-H-R -H-D	Verbal Noun and Prepositional Phrase Verbal Noun and Adverbial Phrase	Appropriate to **B** and **Y**
	Occurrences	Example(s) and explanation	
-H-R	58	*(Dúirt sé liom) - seasamh - ann* = (He told me) - to stand - there	
-H-D	25	*(Dúirt sé liom) - seasamh - go righin* = (He told me) - to stand - straight	
Extension	-F	Pronoun (specific Noun Phrase)	Appropriate to **C**
	Occurrences	Example(s) and explanation	
C-F	62	*Is - é* = It - is (he)	
Extension	A-Z-	Noun Phrase and Particle	Appropriate to **Y**
	Occurrences	Example(s) and explanation	
A-Z-Y-R	49	*Fear - a - bhí - ann* = A man - (that) - it was - there	

5.4 Sentences composed of two clauses

For the purposes of TUIG sentences have been defined as being basically a clause (the Irish word clásal was used since "clause" has a different meaning in PROLOG). However a sentence could also be two "clásal"s joined by a conjunction - defined as a sentence followed by a conjunction followed by a clásal. In all of the cases given in Buntús Gaeilge the second "clásal" would be accepted as a sentence under the definitions given above. This allows TUIG to treat these sentences for parsing purposes as two "clásal"s joined by a conjunction, and thus to re-use the rules already defined.

Conjunctions are of two types: W which takes a dependent form of the verb, and Z which takes an independent form.

Conjunction W

	Occurrences	Example(s) and explanation
-**B**	116	*(Dúirt sé) - nach - mbeadh sé* = (He said) - that - he would not be

Conjunction Z

	Occurrences	Example(s) and explanation
-**B**	191	.....- *a deir sé* = ... (that) - he said

6 Implementation

These rules are applied recursively. This has three main advantages: firstly it is more economical in code and secondly it recognises more sentences than defined by the rules above. For example since D- can in fact be prefixed to any clause - although only D-**B** is quoted in Buntús Gaeilge - extra valid sentences which presumably accounted for less than twenty occurrences each in the study can be included.

However this has the drawback of potentially recognising spurious sentences. To avoid this possibility the implementation of TUIG allows for the brief sentence structure, for example B-A-R-D, to be returned along with the parse structure. These structures can be easily compared with a list of valid structures to eliminate non-valid sentences and this helps to avoid eliminating valid sentences which have not been as yet described.

The third advantage of this approach is that it facilitates that part of TUIG which approximates a Transformational Grammar. In some respects word order in Irish is flexible, as it is in English. The different sentences represented in the example below, showing the different places where the pronoun could occur [ÓS89], would all be recognised and indeed related by TUIG whereas a stricter interpretation would return them as separate sentences.

Fágadh é ina luí ar an talamh taobh thiar den scioból aréir

^ ^ ^ ^ ^

= Was left it on the ground behind the barn last night

It was left on the ground behind the barn last night

This is implemented in TUIG as:

Sentence ---> Sentence + Conjunction + Clásal
Sentence ---> Clásal

Clásal ---> Clásal + Prepositional Phrase
Clásal ---> Clásal + Adverbial Phrase
Clásal ---> Clásal + Noun Phrase
Clásal ---> Adverbial Phrase + Clásal

If Basic Structure is **B** or **Y**

Clásal ---> Clásal + Special Preposition + Verbal Noun
Clásal ---> Clásal + Verbal Noun + Prepositional Phrase
Clásal ---> Clásal + Verbal Noun + Adverbial Expression

If Basic Structure is **C** or **Y**

Clásal ---> Clásal + Adjectival Phrase

If Basic Structure is **Y**

Clásal ---> Noun Phrase + Conjunction + Clásal

7 Examples

Since TUIG is designed to understand analyse sentences relating to the weather, a good test of TUIG would be to attempt to understand a weather forecast. For this purpose a selection of weather forecasts used by RTÉ on *Nuacht* during March 1990 were obtained and the composite text given below as Text 1 is based on these.

As a further test a *seanfhocal* or proverb relating to the effects of various winds was also used. This demonstrates the flexibility of TUIG in understanding sentence types slightly outside the strict boundary of weather forecasts or reports and in the realm of general comment on the effects of the weather. This seanfhocal is given below as Text 2. It is a particularly good choice as it illustrates some of the many uses of the verb *cuir* in Irish, which can have different meanings in different contexts and when used with different prepositions.

Text 1

Beidh ceobhrán san iarthar anocht. Leathfaidh sé soir trasna na tíre i rith na hoíche. Glanfaidh sé amárach agus éireoidh sé gaofar. Beidh an teocht anocht idir 3 chéim agus 7 gcéim.

(There will be drizzle in the west tonight. It will spread east over the country during the night. It will clear tomorrow and become windy. The temperature tonight will be between 3 and 7 degrees.)

Text 2

An ghaoth aduaidh, bíonn sí crua agus cuireann sí fuacht ar dhaoine.
    An ghaoth aneas, bíonn sí tais agus cuireann sí rath ar shíolta.
An ghaoth aniar, bíonn sí fial agus cuireann sí iasc i líonta.
    An ghaoth anoir, bíonn sí tirim agus cuireann sí sioc san oíche.

(The wind from the north is hard, and it makes people cold.
    The wind from the south is damp, and it encourages seeds.
The wind from the west is generous, and it puts fish in nets.
    The wind from the east is dry, and it causes frost in the night.)

The output generated by the first sentence of each of these texts during the various phases of TUIG is given below.

The output of the pre-processing phase of TUIG is usually invisible to the user but is shown below as an example. The input is first organised into tokens. Note, for example, in the first sentence where the preposition and article are separated and *san* is changed to "i","an".

Text 1

token list =

```
[beidh,ceobhrán,i,an,iarthar,anocht]
```

Then the input is tagged, and features which are already apparent at this stage are noted. For instance in the second sentence the mutating "h" has been removed from "oíche", and a "séimhiú" (s) and an "urú" (u) have been identified in the fourth sentence.

tagged list =

```
[(beidh,b),(ceobhrán,b),(i,b),(an,b),(iarthar,b),(anocht,b)]
```

Text 2

The pre-processing of the second text is quite similar. Note how the comma is replaced by "x_camóg", and how the "x" later becomes the tag to signify a conjunction.

token list =

```
[an,ghaoth,aduaidh,x_xcamóg,bíonn,sí,crua,agus,cuireann,sí,fuacht,ar,d
haoine]
```

tagged list =

```
[(an,b),(gaoth,s),(aduaidh,b),(xcamóg,x),(bíonn,b),(sí,b),
   (crua,b),(agus,b),(cuireann,b),(sí,b),(fuacht,b),(ar,b),(daoine,s)]
```

7.1 Parsing

Text 1

In this first example the basic type is "Y", the verb *bí*, and that the sentence has appended in reverse order Adverbial Phrase "D", Prepositional Phrase "R" and Noun Phrase "A". These phrases are then given. The "D" is not repeated in the case of Adverbial Phrases.

```
beidh ceobhrán san iarthar anocht
_x='S'('+D'('+R'('+A'('Y'(
'Y'((bí,f))),
'A'('Ns'(ceobhrán))),
'R'(i,'A'(an,'Ns'(iarthar)))),
anocht))
```

Text 2

The four sentences in Text 2 have the same syntactic structure as can be seen from TUIGs output below. They have of course different meanings, particularly in the prepositional phrases. The most interesting feature of these sentences is that the subject comes before the verb, quite unusual in Irish, but is then repeated as a pronoun in its usual place directly after the verb. This construction was mentioned earlier. This is what the "-JW-Z" is the structures below refers to.

Recall that the explanation of the various symbols used in the parse structure is given above. The sentences are each split into two clauses separated by the conjunction *agus* (and). The first clause is made up of a noun phrase before the verb including the noun *gaoth* (wind) and an adjective, the verb *bí* (to be) and a noun phrase with the pronoun *sí* (she or it) and an adjective. The second clause is made up of the verb *cuir* (to put), a noun phrase containing the pronoun *sí* (she or it), a noun phrase containing the object of the verb, and a prepositional phrase.

```
an ghaoth aduaidh, bíonn sí crua agus cuireann sí fuacht ar dhaoine
_x = 'S'('+A'('-JW-Z'(
'A'(an,'J'('Ns'(gaoth),aduaidh)),
xcamógx,
'Y'('Y'((bí,gl)))),
'A'('J'('NP''PRO'(f3),crua)))),
agus,
```

```
'+R'('+A'('B'(
'B'((cuir,1),
'A'('NP''PRO'(f3)))),
'A'('Ns'(fuacht)),
'R'(ar,'A'('Np'(daoine,duine))))
```

8 Conclusions and Summary

8.1 Summary

Though based on standard natural language processing theory, TUIG joins to this a number of new features which though specifically appropriate to dealing with the Irish language may well prove to have a wider application.

These included, in the first phase, replacing words which had been mutated and inflected, particular properties of words in Irish, with their basic form while maintaining the semantic information available in the actual forms which occurred.

The second phase, parsing, is a standard phase in natural language processing and in the case of TUIG was based around a Phrase Structure Grammar. This formalism allowed a particular property of Irish, that phrase order in the sentence is more flexible than word order within the phrase, to be dealt with efficiently.

8.2 How well did TUIG parse?

TUIG succeeded in parsing all of the common sentence constructions in Irish, and most of the possible ones. It is not surprising then that the sentences in the example texts were parsed successfully. Furthermore TUIG parsed correctly over 90% of the sentences in the weather forecasts provided by RTÉ from which Text 1 was gleaned.

This high success rate is due to the high coverage rate of TUIGs syntactic processing. Since this part of TUIG does not depend on the subject domain except in its vocabulary of nouns and verbs, it is most likely that this success rate could also be achieved in other domains if other domain dictionaries were prepared.

However the syntactic processor is quite a large program, and natural language processing programs require a large amount of evaluation space due to the nature of the processing and the large amount of backtracking involved. This can cause slow execution, and necessitate the trimming of the program. Various implementations of Prolog were tried to choose the most effective and this was found to be Open Prolog, an implementation of Prolog produced within the Department of Computer Science in Trinity College Dublin by Michael Brady.

8.3 Conclusions

The overall conclusion is that TUIG represents a feasible approach to the problem of natural language processing. In addition it is felt that this approach is a useful addition to the field; not only is the approach itself new, but in attempting to analyse Irish it is working on a hitherto neglected part of the field.

This overall conclusion is reinforced by the publication in March 1990 of a paper in IEEE Software which recommended a not dissimilar approach [G&S90]. This approach also used Prolog, and divided the problem into a parsing stage and a semantic analysis phase broadly similar to the division in TUIG.

However in the system described in this paper use was made of a Definite Clause Grammar, which was judged inappropriate for TUIG since it did not allow for the creation of identical semantic representation from different sentences containing the same information. This feature of TUIG has already been described, and the authors of this paper admit the drawback with their approach.

The second conclusion, that this approach not only adds to the study of natural language processing but adds to a part of that study which had not yet been undertaken to any great extent, is no less important. Some of the described features of TUIG are unusual, such as the "Tagging" and coding of sentence structures, and it is hoped that they might find a wider use in the field. The absence of significant work on Natural Language Processing in Irish meant that much preliminary work had to be done, but also that TUIG was not constrained by having to reflect or compete with other systems.

References

[G&S90] Geetha, T.V., and Subramanian, R.K., "*Representing Natural Language with Prolog*" in IEEE Software March 1990

[MGPh63] Mac Giolla Phádraig, B., *Réchúrsa Gramadaí (3ú eagrán)*. Longman Brún agus Ó Nualláin, Baile Átha Cliath

[ÓH66] Ó Huallacháin, C., *Buntús Gaeilge*: réamhthuarascáil ar thaighde teangeolaíochta a rinneadh sa Teanglann, Rinn Mhic Gormáin. Rialtas na hÉireann, Baile Átha Cliath

[ÓS89] Ó Siadhail, M., *Modern Irish*: Grammatical structure and dialectal variation, Cambridge University Press

Section 3:

Knowledge Acquisition

Information Retrieval

User Modelling

Knowledge Acquisition in the Engineering Domain : A Case Study

Maureen Murphy, John G. Hughes and Michael McTear
Dept. of Information Systems
University of Ulster at Jordanstown
Newtownabbey
Co. Antrim BT37 OQB
N. Ireland

Abstract

In aerodynamics technology, the transfer of basic research to the design engineering community has long been recognised as a problematic and time-consuming task. Thus the idea of designing an expert system to act as an intermediary, facilitating communication between these two groups was engendered. Faced with the problem of knowledge acquisition in this highly technical domain, a novel approach was developed. This paper introduces and investigates the effectiveness of this technique which employed the skills of a "Knowledge Interrogator" in the acquisition process.

Introduction

In aerodynamics technology, the transfer of research results to the design engineering community is critical for the timely advancement of new aerospace designs. Researchers currently make public their work through the medium of technical journals and presentations at conferences. This practice presents a two-fold problem to the design engineer anxious to keep abreast of advancements within the field :-

1. The vast quantity of printed literature necessitates long hours filtering through publications often irrelevant to design objectives.

2. Research papers are frequently written in a genre and phraseology intended to be read and understood by fellow researchers only.

A need exists therefore for an intermediary, familiar with both research and design to provide designers with research knowledge that is understandable and relevant to their needs. Availability and cost considerations often mean that human experts cannot be employed, thus the idea of designing an expert system for the task was engendered.

Complications and complexities can present themselves during the development of an expert system in any domain but problems peculiar to this application were recognised as being significant from its inception. These included :-

1. The volume of knowledge

2. The accuracy of the human expert's biases in deciding which basic research results are relevant.

3. The necessity of updating the knowledge base frequently in order to keep pace with research results.

4. The proposed structure of the knowledge base as one of the aims of the project was to design a 'skeleton knowledge base structure' that could be applied to the transfer of any subset of aerodynamics technology.

5. The inadequacy and inappropriateness of the techniques currently employed for knowledge acquisition for the elicitation of expertise within areas of advanced technology.

This paper concentrates on the latter point, highlighting the inadequacies of current knowledge acquisition techniques for use in this highly technical domain and introduces and investigates the feasibility of an innovatory technique employed during the acquisition stage of the knowledge engineering task.

The scope of the system was limited to the design and fault diagnosis of a low-speed wind tunnel for use by both design engineers and engineering managers working under cost and technical constraints.

Wind Tunnel Objectives

The aim of the knowledge-based system is to help a design engineer or manager answer the following questions about a potential wind tunnel design :-

1. Will the wind tunnel be satisfactory given the desired features and the perform-

ance and practical constraints?

2. What is the accuracy of the data that will be taken in this tunnel?

3. What will be the difficulties involved in building and maintaining this tunnel?

4. What additional information is needed to build this tunnel?

5. What is the cost of this tunnel, including location and staff?

In addition to the general questions above, the system must also deal with very detailed information on the physics of the flow field within individual wind tunnel components. This information would allow a design engineer to access research knowledge which is timely and relevant to his/her design interests.

In order to address the needs of novice design engineers, it was proposed that the system also make available on request a tutorial providing 'textbook information' on terms and concepts encountered during a consultation.

Knowledge Acquisition

During the primary stages of development it was hoped that a time-effective, cost-effective and knowledge-intensive elicitation process could be carried out. Experience has shown that this is not often achieved as problems frequently encountered include :-

1. Domain specific terminology and concepts which the knowledge engineer must be able to comprehend

2. Problems relating to the ability of experts to express their knowledge

3. Difficulties inherent in planning an organised approach to knowledge acquisition

4. Selection of appropriate and effective knowledge acquisition techniques

The current techniques employed were investigated and reviewed for their applicability to the capture of knowledge in this domain, some of which are listed below :-

1. Text Analysis : Knowledge is acquired through the use of text-books and

manuals without involvement with a domain expert.

2. Behavioural Analysis : The knowledge engineer makes observational studies of the expert at work allowing him/her to examine discrete areas and compare what the expert says with what is done.

3. Machine Induction : A general rule is induced from a set of example cases.

4. Interview Analysis : This incorporates the use of semi-structured and structured interviews, verbal protocols, task analysis, multi-dimensional scaling, repertory grids and card sorts.

Text and behavioural analysis were discounted as being inappropriate for the diagnosis of a wind tunnel design as observational studies of an expert at work could not be carried out and involvement with experts in the field was deemed imperative.

Rule induction techniques or the use of knowledge acquisition tools [7,14], although having gained some limited practical success in industry, were deemed to be incapable of coping with the complexity of the domain under study.

Research in the field of knowledge acquisition has in the past tended to focus on interview analysis [1,2,4,5,6,8,12,13]. In practice repertory grids and card sorts provide a good means for knowledge engineers to elicit an expert's conceptualisation of the domain . With regard to protocol analysis, however, Burton et al [1] found that during their evaluation of four acquisition techniques, (interviews, protocol analysis, repertory grids, and card sorts) protocol analysis proved the least satisfactory in terms of time and analysis with a smaller amount of necessary knowledge gleaned. Although the benefit of these techniques was recognised, both in terms of domain conceptualisation and elucidation, their use was again found to be inappropriate for the acquisition of design knowledge in this complex knowledge intensive domain. The inherent concepts of aerodynamics are not only complex but are also shrouded in a terminology which would in itself take an inordinate time for a knowledge engineer to become familiar with.

During the elicitation stage, interviews have always played a leading role though their success is heavily dependent upon the interviewing skills of the knowledge engineer, his/her familiarisation with the domain and the attitude and responsiveness of the domain expert(s). The major stumbling block with this acquisition technique did not lie with the attitude of the experts but again with the very technical nature of the application domain. Traditionally a period of orientation is recommended for the knowledge engineer before commencing an interview session in order to 'alleviate' the problem of domain incomprehension, as Prerau [11] notes,

> " It is useful to invest some time up front discussing the domain in general without focusing on the task the expert system will do."

These discussions can take the form of open-ended interviews with the domain expert or more formal domain-related training seminars presented to the knowledge engineer. Further measures include reviewing company documentation, observation and studying basic textbooks and manuals. Once familiar with basic terminology and sources of reference, communication is deemed to be enhanced between the knowledge engineer and domain expert. The effectiveness of this sometimes lengthy period was put into question, however, with respect to a domain as complex as that of aerodynamics. Although perhaps familiar with concepts such as 'turbulence', 'jet flows', 'unsteady separated flow', it was thought to be naive to believe that a primarily nescient knowledge engineer would be able to conduct a coherent and constructive knowledge acquisition session within a suitable timeframe. Thus it became apparent that an engineers ability to direct a successful acquisition session, delving beyond 'shallow reasoning' into the deeper knowledge more relevant to the needs of design engineers, was severely limited. Having eliminated current acquisition techniques as inappropriate and inadequate for the task, a new form of knowledge acquisition was proposed, employing the skills of an intermediary knowledge interrogator familiar with the domain during the interview sessions.

On a wider perspective, it was recognised that the use of an interrogator might overcome a problem common in the elicitation stage, poor intercommunication skills on the part of the knowledge engineer. As knowledge engineering is still a relatively new discipline, many practising knowledge engineers have crossed from the field of computer science and are inexperienced in the use of effective knowledge acquisition techniques. The problem of domain incomprehension can sometimes be compounded therefore by 'poorly chosen and delivered techniques'[6] as engineers are untrained in these skills which are borrowed from fields as diverse as communication (eg consensus decision making), psychology (eg construct analysis), instructional design and education (eg lecture). As a prerequisite in the choice of interrogator was managerial and interviewing skills, it was hoped that, where relevant to the nature of the domain and the intercommunication skills of the engineer, the use of an interrogator, would remove this potential impediment to the acquisition phase of development.

Knowledge Interrogator

The role of the interrogator is to act as a liaison between the user, knowledge engineer and the expert. As an individual familiar with the application domain, possessing both managerial and interviewing skills whilst also aware of the needs of the end-user, his/her responsibilities exceed those of a 'knowledge co- ordinator' [6] who,

1. ensures that the knowledge engineer's domain familiarisation activities proceed as required

2. reviews the knowledge acquisition plans proposed by the knowledge engineer.

3. sets up the knowledge acquisition sessions upon the knowledge engineer's request

4. monitors the knowledge acquisition session completion and effectiveness

5. oversees the translation of the knowledge from source to the initial code

The interrogator plays more than an organisational / managerial role throughout the acquisition period performing tasks conventionally carried out by the knowledge engineer. Responsibilities assumed include choosing the domain experts, scheduling and drawing up the agenda for the proposed interview sessions, and most importantly leading the questioning of the experts.

Throughout the interviewing sessions the knowledge engineer plays a passive role, the onus only being placed on him/her during the knowledge representation and encoding stages.

The Wind Tunnel Project

During the wind tunnel project the primary duties of the interrogator were as follows :-

1. **Choice of Experts** : The interrogator, with a wide circle of intellectual acquaintances, chose individuals whom he believed would prove most knowledgeable and experienced in the field for the collection of scientific facts and expertise.

2. **Agenda** : As an individual technically skilled in the general area of the domain, while also familiar with the needs of end-users, the agenda for the knowledge acquisition sessions was drawn up by the interrogator with help from a domain expert. The agenda was distributed to the participating experts before the first knowledge acquisition session to allow the domain experts to

make preparatory notes, detail possible amendments and seek references and knowledge sources.

Knowledge Acquisition Session

The participants in the knowledge acquisition session included :-

1. **Interrogator** : technically skilled in the application domain with managerial, interview and technical skills

2. **Experts** : three experts were used for the knowledge acquisition sessions

3. **Knowledge Engineer** : familiar with some of the more common aerodynamic terminology and concepts.

It was decided that the interview sessions should take place in the office of the principal expert to allow for easy access to references without any undue interruption to the session. Before commencing the acquisition session the experts were given the opportunity to make amendments or additions to the proposed agenda recognizing that the interrogator, whilst familiar with the general domain was by no means infallible in his choice of topics of interest.

A semi-structured interview technique was adopted which, as Hoffman [15] states, "forces the domain expert to be systematic". The session began with the first item on the agenda with questions directed to the first expert and then the second for their agreement, comments and expansions on any points. The agenda was methodically worked through with the interrogator ensuring that the experts did not diverge too greatly from the outline structure and provided knowledge and expertise which would be of benefit to the resulting knowledge base. Aware that an expert system knowledge base is not solely concerned with scientific facts and figures, the interrogator encouraged the experts to articulate their own personal heuristics developed throughout their years as these practices often carry more weight than any specific analytic technique within the industry.

Using techniques such as

paraphrasing, "What I believe you said was..., in other words..."

clarifying, "Please repeat that for me ..."

summarizing, "To sum up what you have been saying..."

the interrogator was able to conduct three 2-hr acquisition sessions which resulted in the extraction of over 300 rules in the resultant knowledge base.

One important aspect of this technique was the use of audio-equipment replacing the standard practice of copious note-taking by the knowledge engineer. The continuity of the session was subsequently increased as neither the interrogator nor knowledge engineer were occupied with detailing information but were able to focus on the direction and content of the conversation more effectively.

The session was later transcribed for use by the knowledge engineer during the knowledge representation and coding stages. The transcript was subsequently edited to a readable form with common topics placed together, forming the basis for the tutorial included in the wind tunnel knowledge base.

Knowledge Representation

The duties of the interrogator did not end with the elicitation stage but continued through to the next phase of knowledge representation.

On receipt of the typed transcript, the knowledge engineer was able to formulate over 100 rules. These were given to the interrogator for review, who, once familiar with the representation technique for the knowledge base, was able to analyse the rules constructed by the knowledge engineer and provide suggestions for additional production rules.These new items of information were then incorporated into the knowledge base, after some manipulation, almost doubling its size. [Appendix I]

Following the ethos of prototype development, the wind tunnel knowledge base was constructed incrementally. At each stage the participating experts were given the opportunity to express their comments and provide constructive criticisms.

In order to aid the debugging of the knowledge base a series of test cases was constructed and the results and advice produced by the system compared with those given by a human expert.

Wind Tunnel User Interface

It was determined that the expert system user interface should be in the form of an interactive query session with the user detailing the characteristics of a proposed preliminary wind tunnel design. Unsatisfactory features would then be highlighted to the engineer with the option to review knowledge of the relevant fluid mechanics in greater detail.
Presented with a main menu of seven options, the user can for example 'Do a complete Analysis', 'Check for General Problems', 'Check Cost Factors', or detail a specific com-

ponent.

During a consultation the system requests information on various aspects of a preliminary design including room conditions, the test section dimensions, power and labour requirements and individual tunnel features such as the maximum and minimum tunnel speed and accuracy for the flow angularity. Advice provided by the system falls into four categories :-

1. General problems : "The vibration level may be too high with fibreglass."

2. Recommendations : "Water is recommended as the working fluid when flow visualisation is required."

3. Cost considerations : "The cost of maintaining screens may be too high due to restricted access."

4. Pathological problems : "The diffuser design may cause flow unsteadiness, please consult an expert designer."

A second knowledge-based system was constructed using similar acquisition techniques as those employed by its parent wind tunnel system. Although less elaborate than those built by Lo and Shi [10], this system allows engineers to access a wide body of structured knowledge dealing with the underlying fluid mechanics of one component of a wind tunnel, the diffuser.

Conclusion

The success of this approach to knowledge acquisition is of course highly dependent on the choice of Knowledge Interrogator. An ideal candidate would be an individual who possessed communication, interview and technical skills. The latter, technical skills, must encompass knowledge of the application domain, though not necessarily expertise, and an appreciation of expert system technology. An interrogator must also be aware of the needs of end-users in order that the choice of subject matter for the relevant agendas be correct and complete.

It could be argued that the problems encountered by a knowledge engineer during the elicitation stage have been moved rather than removed by the introduction of a new component. The Knowledge Interrogator assumes the traditional problem-ridden responsibilities of choosing the domain experts, eliciting the knowledge and helping to collate the information gleaned. Realistically, do such multi-talented individuals exist with the time to carry out these tasks? Is it not an additional link in an already weak chain?

It is not proposed here that a Knowledge Interrogator is the panacea to all acquisition difficulties. It is recognised that an ideal interrogator, displaying the desirable qualities detailed above may not be available, nor be required, in all proposed applications.

During the wind tunnel project the system developers had the foresight to realise that a worth-while system could not be developed within a cost-effective and time-effective period by an engineer outside the field. Another option proposed [6] is that a domain engineer becomes proficient enough in the field of expert systems to build the system him/herself. Again this was rejected on the basis of the time and effort which would need to be expended by the domain experts who could not be regarded as 'computer-literate'.

Within the group an individual did exist who was familiar with the domain and had sufficient experience in interviewing techniques to conduct an acquisition session and on these grounds was chosen as the interrogator. He was not familiar however with expert system technology. As only the fundamentals are required, the actual construction of the knowledge base being still within the duties of the knowledge engineer, an introduction to the field of expert systems was undertaken and completed within a short period. This task was considered a trivial one in comparison to the education that would be required by a knowledge engineer unfamiliar with the field of wind tunnels and fluid mechanics.

Thus within this field of advanced technology the employment of a Knowledge Interrogator led to the development of a knowledge-based system within a time and cost effective period. The resultant system captured detailed information and knowledge and presented it in a comprehensible and coherent manner and was approved by the domain experts as being a useful tool for practising engineers in the field today.

References

[1] Burton A.M., Shadbolt N.R., Hedgecock A.P. & Rugg G. A formal evaluation of knowledge elicitation techniques for expert systems : domain 1. In: BCS SGES Research and Development in Expert Systems IV, 1988, pp 136-145.

[2] Hayes-Roth F, Klahr P, Mostow D.J. Knowledge Acquisition, Knowledge Refinement and Knowledge Refinement, In: Klahr P, Waterman D (eds), Expert Systems: Techniques, Tools and Applications. Addison-Wesley Publishing Company, 1986 .

[3] Quinlan J.R. Semi-autonomous acquisition of pattern-based knowledge. In: D. Michie (ed) Introductory Readings in Expert Systems. Gordon & Breach Science Publishers, 1982.

[4] Richard Forsyth (ed), Expert System : Principles and case studies. Chapman and Hall Ltd, 1989.

[5] Graham I, Jones P. Expert Systems : Knowledge Uncertainty and Decision. Chapman and Hall Ltd, 1988.

[6] McGraw K, Harbison-Briggs K. Knowledge Acquisition : Principles and Guidelines. Prentice Hall Inc, 1989.

[7] Eshelman L, McDermott J. MOLE : A knowledge acquisition tool that uses its head. AAAI, 1986, pp 950-955.

[8] Kline P, Dolins S. Problem features that influence the design of expert systems. AAAI, 1986, pp 956-962.

[9] Rodman l, Nixon D, Canning T, Hughes J, Bradshaw P. The use of knowledge-based systems for aerodynamics technology transfer. AIAA, 1991, Num:0500.

[10] Ching F. Lo, George Z. Shi. Development of an Intelligent Hypertext System for Wind Tunnel Testing. AIAA, 1991, Num:0654.

[11] Prerau D. Knowledge acquisition in the development of a large expert system. AI Magazine 1987 pp 43- 52.

[12] Hayes-Roth, Waterman D.A, Lenat D.B (eds). Building Expert Systems. Addison-Wesley Publishing Company Inc, 1983.

[13] Cleal D.M, Heaton N.O. Knowledge-Based Systems : implications for human-computer interfaces. Ellis Horwood Ltd, 1988.

[14] Boose J, Gaines B. Knowledge Acquisition Tools for Expert Systems, Knowledge-Based Systems, vol 2 , Academic Press, 1988.

[15] Hoffman R. The problem of extracting the knowledge of experts from the perspective of experimental psychology. AI Magazine, Summer 1987.

Appendix I - Extract From The Raw Transcript

Expert : Occasionally people have fillets in thecorners of a contraction so that it's not exactly rectangular, but an octagonal shape. The idea of that is that separation is most likely to occur in the corners where two boundary layers meet, and this would reduce that. I think that the most important feature in contraction design is to make sure that the typical radius of curvature in side view or top view at the wide end, is fairly large, with the radius of curvature at the bottom end, at the downstream end being comparatively small. This seems to be the best way of winding up with a nearly uniform flow at the start of the working section. If you try to make the contraction too short, or if the shape near the downstream end is inadequately chosen, then you can get distortion of the streamlines in the first part of the working section because the flow is still trying to converge after you get into the working section itself. It's not possible to get exactly uniform flow coming out of the end of the contraction into the working section, potential flow theory just forbids this.

Interrogator : From a scientific point of view, the words " seems to be the best way" have a certain lack of precision.

Expert : This is our subject in general. It's partly based on practice and I think that most people would do would be to take an existing wind tunnel contraction shape and adjust it slightly for their own purposes.

Interrogator : Is there any scientific basis for actually designing a contraction? Seems to me the best way is very imprecise. It seems strange in this day and age of computer programs that can duplicate entirely wind tunnel results, especially the errors,that we cannot, in fact, find a better way to design a contraction.

Expert : You've mentioned a way of analysing the flow in a given contraction, so you have to choose your shape and if the contraction design appears to be critical, you can either build a model to test it, or do a calculation to test it. It's a non-trivial calculation, you'd have to do out a complete Navier-Stokes calculation or a combined boundary layer plus inviscid flow calculation.

Expert : It's not a very critical area of wind tunnel design. It takes a fair amount of incompetence to get separation in a converging passage anyhow.

Extract From The Rules Derived from the Transcript

IF contraction ratio <= 7 and >=12
AND contraction properties include "length is 1 to 2 times maximum diameter" and "radius of curvature large at inlet, small at outlet"
THEN contraction is satisfactory

IF contraction is not satisfactory
THEN general problems include "The contraction design is unsatisfactory"
AND explanation is "The contraction should be between 7 and 12, and the contraction should have the characteristics listed on the form"
AND recommendations include "The contraction is close to design limits - consult an expert designer"
AND references include 8 and 9

IF critical factors include test section boundary layer
THEN recommendations include "Consider using panel/boundary layer codes to compute boundary layers in contraction"
ELSE recommendations include "Use existing design for contraction"
AND explanation is "If the test section boundary layer is not a critical factor, then a conventional design can be used for the contraction such as a 5th order polynomial"

Intelligent Searching through Hypertext Systems for Learning

Alan F. Smeaton

Catherine Guinan[1]

School of Computer Applications

Dublin City University

Glasnevin, Dublin 9.

Abstract.

One of the most attractive ways of accessing multimedia and even simply textual information in the last few years has been to use hypertext techniques. In this paper we discuss one of the application areas of hypertext systems, namely computer-assisted learning (CAL). Learning as a cognitive process is briefly reviewed, as are current approaches to using computers for instruction, i.e. intelligent tutoring and computer based training (CBT). We then discuss the current approaches to using hypertext techniques in CAL and discuss some of their inadequacies. This leads us on to an outline of the work we are doing to enhance the functionality of hypertext delivery systems by providing them with the capability to dynamically create *guided tours* through the hypertext, in response to individual student or user needs. Such an enhancement combines the discipline of CBT, the individual tailoring of intelligent tutoring and the freedom of hypertext browsing. Our plans for implementing such a facility are presented.

---

[1] The second author was funded by Telecom Eireann and the Irish-American Partnership Scholarship Fund.

1. Introduction

Multimedia and especially textual information is becoming increasingly commonplace on todays computing systems. More and more we are storing text, images, photographs, graphics, animation, sound, voice, etc, mostly for applications like document production. Since effective content-based retrieval on most forms of non-traditional data is at least a very difficult process, alternative ways of accessing multimedia information have been proposed. One of the most attractive ways of accessing multimedia and even simply textual information, has been to use hypertext techniques.

Hypertext is computer-supported non-linear viewing of information where it is the reader or browser who can choose to view information in any order desired. A hypertext information space consists of a collection of *nodes* or chunks of information which each represent independent and autonomous fragments of the overall information content of the hypertext. Nodes can contain information in any media. Each node is linked to a set of other nodes in the hypertext via specially authored *information links*. A user reads a hypertext by *jumping into the hypertext* using a simple search method, and then browses around the hypertext from node to node until the original information need has been satisfied. The user-controlled *browsing* of an information space means the user has the choice of which information links to follow.

Although hypertext has only emerged as a popular way of organising information within the last 3 to 4 years, the ideas behind it go back a few decades. Hypertext is only a way of organising information and within the very recent past personal computing technology has developed enough to allow a profusion of hypertext and hypermedia development systems and products to emerge onto the marketplace. Such systems allow users to create their own hypertexts using an authoring tool, and they also allow browsing through the created hypertext using a separate browsing tool. Such hypertext development systems are available for a range of computing platforms, from lowend personal computers, to UNIX-based workstations. In addition to development kits, a number of vendors are also selling ready-made hypertexts where the information has already been created. In essence this involves the selling of information which has been organised in hypertext form. Typically this information would be in read-only form, often distributed on a CD-ROM disk.

Applications of hypertext, both ready-made and user-created, are now quite numerous and varied in nature and include online documentation, software engineering, dictionaries, tourist information and learning. Most major computer manufacturers including DIGITAL, SUN, Hewlett Packard and SYMBOLICS distribute reference material in machine-readable format where the information has been organised into hypertext form. For the software engineering area the hypertext information space consists of information related to the development and maintenance of a software system including requirements specifications, program code, notices of bugs and fixes, user documentation, etc., all judiciously linked with hypertext information links [GARG90]. The Oxford English dictionary is being converted into hypertext format and will be distributed on CDROM [RAYM88]. In the area of learning there have been literally thousands of developments of hypertext systems for learning/teaching. For example at Harvard University the PERSEUS Project is providing hypertext support for the study of ancient Greek literature, history and architecture [MYLO90]. At Dublin City University a hypertext has been developed as adjunct courseware for an undergraduate course in databases [SMEA90, SMEA91].

Because of the surge of interest in using hypertext technology quite a few problems inherent with hypertext organisation have surfaced and one problem which is present in ordinary linear text but which is accentuated when placed in the hypertext concept is one of user disorientation. This is the problem of the user not knowing where they are in the hypertext network and also not knowing how to get to somewhere that they know exists elsewhere in the hypertext. In traditional linear text, the readers always know exactly where they are and in order to find some other information they only have the choice of searching forwards or backwards. Because of the multi-dimensional aspect of hypertext, readers have many more options available if they want to locate some information and this can prove to be very confusing. The user needs to feel comfortable with the system. If they decide to follow a certain path of information links, then it should be possible for them to retrace their steps or at least return to the starting point. If the system leaves the user with the impression that they will become lost if they follow a certain link, then it will be useless as no-one will utilise it. The most commonly proposed solutions to the disorientation problem include graphical browser maps [HALA88, EGAN88, CONK88] and specifically authored guided tours [TRIG88].

Another problem intrinsic to hypertext is the cognitive overload involved in using it. Both authors and readers of hypertexts experience the problem of having too much information on hand at certain times by virtue of the information links. When authors are creating hypertexts, several ideas are occurring to them simultaneously and unless a note is taken of these ideas, they may vanish. Unfortunately, the effort required to leave the current task, open a new node, create and type links, etc., may prove too much for many authors and they will just become over-burdened with all of this information they have to juggle. This problem is not really specific to hypertext systems as authors of any documents experience the same problem of having to keep track of many ideas at the same time but like disorientation, it is accentuated when dealing with hypertext. Almost all of the many suggested ways of reducing the cognitive strain on hypertext users are to do with easing the disorientation problem but this is treating the symptoms rather than the cause. It may be that the nature of the intellectual strain on hypertext users is not really understood, and therefore cannot be eased. This is a point we shall return to later.

One other serious problem with using hypertext systems is finding a starting node from which to commence browsing and some of the most important work on finding the most suitable starting point is being carried out by Mark Frisse at Washington University School of Medicine. His research is concerned with information retrieval rather than learning from a medical handbook [FRIS88]. The use of indexing techniques from information retrieval research in Frisse's work reduces the burden on the user to find useful nodes by browsing through the hypertext. Frisse has developed a global access method to the hypertext so that if the hypertext is very large the user does not have to spend a lot of time navigating. Instead all of the relevant nodes in the network will be retrieved and the user can select the ones which they deem to be most appropriate to the original query as a commencement to browsing. In small domains this approach can be quite effective as the number of cards selected tends to be rather small. However in large hypertexts the number of nodes returned can be too great to afford the user any savings in time and effort. Frisse, therefore, does not rely solely on pattern matching between a user's query and node contents. His approach considers node size, content, context and link semantics when determining which nodes should be returned to the user as relevant. He has developed methods to increase the probability of selecting useful starting point nodes using statistical frequencies of index terms which enables the implemented hypertext system to take advantage of frequency information with respect to word occurrence and to propagate this information throughout the hierarchically-structured hypertext. This has been shown to provide better nodes from which users can browse the hypertext.

In addition to those problems of disorientation, cognitive overhead and starting locations for browsing mentioned above there are other difficulties which include converting hierarchically structured text into hypertext form, versioning of nodes and links in a hypertext, extensibility and tailorability of systems and the evolution and enforcement of standards for hypertext design and representation. These, however, are more syntactic when compared to the more fundamental problems mentioned above.

One of the applications of hypertext that we are interested in is using it for learning or computer based training. As mentioned earlier there have been many projects which have done this with varying degrees of success. Each attempt has been different as there is no logical way to use a hypertext for learning and in doing so make it as effective and comfortable for the student as far as the learning is concerned. The reason why this is an empirical task rather than one which has an obvious method is that learning itself is not really understood. In our work what we want to do is to explore the scope of hypertext as a learning tool. We have some experience of hypertext and learning based on the hypertext for databases that was created and we have had feedback on its usage from the students involved. Based on our experiences we can recognise the inadequacies of current hypertext products with respect to their suitability as tools for learning and we have some suggestions as to how hypertext systems could be improved in this direction.

The remainder of this paper is organised as follows. In the following section we give a brief overview of what learning is as a cognitive process in general and we discuss some of its features. This is followed by a review of how computers have been used in the learning process in other work using both conventional CAL and intelligent tutoring approaches. In section 4 we discuss the prospects of learning from hypertexts, the weakness of current systems and approaches and the attractiveness of the combination of hypertext and learning, if it can be made to work. Following that we present some details of our specific contribution to moulding hypertext functionality towards a better support of the learning process by the use of dynamically constructed guided tours which plan routes through the hypertext information space. Finally, our conclusions will present some of our plans for our future work.

2. What is Learning ?

Human learning is a cognitive process that comes naturally to us but is proving to be almost impossible to automate. Machine learning as a sub-branch of artificial intelligence has been developing for decades [BRAT88] and has now gotten to the stage where machine learning techniques can be used to automate some kinds of knowledge acquisition in expert systems by inducing rules from sample data [GAIN88].

But what about the process of learning itself. Do we know anything about how the human brain actually learns things ? If we did know about learning then we could use this in some way to devise computing systems that are effective to learn from. In order to make progress what we must do is devise a *cognitive model of learning* which accurately maps onto the way we learn. From this model of learning we could then develop computing systems which are close to or take advantage of how we learn. However there is no agreed cognitive model of learning that we can use, but we do know about some of the characteristics of human learning and we can take advantage of these. This is not quite the same thing as modelling the entire process, but it is progress at least.

Concerning learning we know that there are different depths of learning which vary from surface recognition or skimming, to a deep understanding of concepts. We know that learning is highly contextualised and cognitive studies have shown that what is learned is often tightly bound to the context and situation in which the learning has taken place [LAVE88]. For example when teaching a course on databases one generally uses examples to illustrate examples of SQL query language constructs or aspects of the database catalog. These same examples as used in lectures quite often appear on examination scripts as answers to questions. Perhaps this could be attributed to rote learning or learning-by-heart or perhaps it is an example of contextualisation, we don't know.

Acquiring a deep understanding of a concept needs active involvement of the learner in solving problems related to that concept. This is a technique used extensively in manual teaching at all levels. For example, when teaching a course on databases to University students, we teach the SQL language by getting students to turn natural language statements into SQL. In such a case there is no substitute for practice.

Other aspects of learning that are relevant to the relationship between learning and hypertext are the ideas of *incidental learning* and *discovery learning*. Both of these are independent of the method of learning and can occur with either paper-based or computer-based learning. Incidental learning involves the student learning what is presented to him, either rote learning or learning the abstract concepts above the specific material presented. For example, a student learning details of a relational database catalog might learn just the catalog tables for the INFORMIX database (rote learning) or might learn the typical information which would be stored in a relational DBMS catalog, with the INFORMIX database as an example (learning abstract concepts). Discovery learning occurs where a student learns material which has not been explicitly presented but which has been discovered or created independently by the student. This type of learning does not occur as often as incidental learning, but where it does it leads to a very deep understanding of the material. If a student has discovered something themselves then it is much easier to remember and recall later, than if it has been learnt from presentation. An example of discovery learning would be where a student might learn about the concepts and role of deductive databases, without reading material on their implementation, and might also learn about logic programming and PROLOG, and might independently discover for themselves that deductive databases are suited to implementation using PROLOG, despite the fact that this piece of information is already well-known, but was never previously seen by the student.

From the above we can see that we don't really know a lot about learning, but we know what it should be like. In the next section we shall see how computers and computing systems have already been used to support learning in some sense.

3. Computers and Learning

Computers have been used for teaching purposes for many decades, from the early, rigid computer based training of the 1950's to the more sophisticated tutoring systems of today. The use of the computer for educational purposes has many practical advantages in that students have been shown to progress more rapidly when they receive one-to-one tuition as opposed to being just one of a number of pupils in a classroom situation. Obviously there exist economic reasons why every student cannot be furnished with a private tutor and this is where computers fill the gap. If a system can be developed which teaches the student in the same manner as a human tutor would, then each individual can reach their full potential without being impeded by the rate of progress of their fellow classmates. This is not a trivial task. Much research has been carried out on the cognitive processes of individuals in order to decide how best to design the "computer coach". As we saw when looking at learning, the best way for a student to retain information is to promote discovery learning where students can form their own ideas and concepts from presented material.

Computer Aided Learning (CAL) is understood, in this paper, to have two forms. The first and earlier form is that of programmed instruction which is closest to our idea of computer based training (CBT). The second form is that of Intelligent Tutoring Systems (ITSs) whereby artificial intelligence techniques are incorporated into education in order to produce tutoring systems which more closely resemble their human counterparts. Both these approaches have attempted to overcome the difficulties inherent in using the computer as a teaching medium, albeit in different ways. The most significant problem to be overcome by designers and developers of such systems is the fact that all learners have different rates of learning, prefer different approaches to learning, have different objectives, are motivated differently, and so on. For a student to gain maximum benefit from using the computer as a teaching mechanism, the teaching system must be such that it adapts to the different needs and styles of the various users.

3.1 Programmed Instruction.

This was the first major effort to find a solution to the aforementioned problem of different learning styles and rates among different users. Previously, the computer had been treated as little more than a programmed text book in that all students were presented with the same information in the same sequence. These systems did not provide any degree of individualisation and no account was taken of the users reaction to the material. Vast improvements were made with the development of programmed courses which were designed for specific categories of learners. These courses are interactive and develop new knowledge and skills in students through a sequence of exercises, presented so as to provide each individual with a great deal of proficiency in the area being studied.

The author of these programmed instruction courses must pay attention to three fundamental principles of course design [DEAN83]. Firstly, it is important that they distinguish between leading the students step by step through the information and leaving them free to roam around at will. If they are left to their own devices there will probably be a high degree of incidental learning and the students will more than likely have a greater understanding of the studied material because they have been allowed to form their own impressions rather than have the designers opinions forced upon them. However, they run the risk of not studying all of the available information and may therefore lose out on some significant aspects of the material. Effective courses tend to include frequent questions and problems testing understanding at successive stages of learning. Thus, the students attention is brought to areas which may not have been fully understood. Secondly, the system should provide the student with some feedback as to how well the material is being understood from the answers being given to questions on the material. Finally, the structure of the teaching process should be such that the student gradually progresses towards proficiency on the subject matter. If the material is poorly designed, the student will end up turning in circles instead of directing their attention towards a specific end.

To provide some degree of individualisation and learner control, remedial sequences of frames for all incorrect responses must be inserted. This results in the author providing a complex maze of paths through the information. The danger remains, however, that if a student selects the correct response either by skill or by chance, thereby by-passing the remedial or explanatory frame, they may lose out on some valuable information contained within these frames which other students explored. The author must ensure that all important information is contained in the main structure and that the remedial frames consist only of supplementary material which would assist but not significantly improve the knowledge of any student who finds it necessary to study the material contained within.

The Centre for New Technology in Education is a self-funding unit within Dublin City University which provides a service of producing CAL packages for use within and outside the University. The systems which they use to produce these packages are SAM and TEN-CORE. They have produced training systems for external organisations and also a package which simultaneously trains the learner in the use of a French banking system and the French language. The structure of this French package is very sequential, the user having to complete one block before proceeding to the next. At the end of each block, the user is questioned to see if they have understood the material thus far. The structure of the package is not rigidly linear; some deviations are allowed by the user. However, when each indivdual has completed the course, they are all at the same degree of proficiency. Much of the product development time is taken up with developing a suitable user interface. It is important that the package should be easy to use and does not frustrate the student. Alongside the user interface, *storyboarding* is an important activity. This is the idea of designing routes through the network, deciding what questions to ask and how to phrase them, etc.

The work of this unit at Dublin City University is typical of the state of the art in CBT and their success as a venture shows the achievements of this approach to programmed instruction.

3.2 Intelligent Tutoring Systems.

Intelligent tutoring systems are computer programs that are designed to incorporate artificial intelligence techniques within educational applications in order to provide tutors which know what they teach, who they teach and how to teach what they teach [NWAN90]. Designers of such systems must take into account that ITSs lie in the intersection of three disciplines, namely, computer science, psychology and education. Therefore, an understanding of all three areas must be attained before a true ITS can be developed. ITSs overcome the individualisation problem by providing immediate feedback to the user. Tutoring is thus more effective as it occurs in direct response to the needs of the student.

ITSs differ widely in their architecture however there is a general consensus that they consist of at least four basic components :

1. the expert knowledge module
2. the student model module
3. the tutoring module
4. the user interface module

1. The *expert knowledge module* consists of the facts and rules of the particular domain to be conveyed to the student i.e. the knowledge of the experts. Acquiring this knowledge and presenting it in a suitable format is a very time-consuming task and is the central activity in the creation of this type of module. In current ITSs, expert knowledge is represented by means of semantic networks, frames, production systems and others. Not only must the knowledge be represented in a form which facilitates explicit learning but also in a manner whereby an implicit understanding of the representation may be formed. This has more to do with the structure of the material rather than the content and is, as such, much less specific and therefore more difficult. This expert knowledge module fulfils a double function. The first, we have seen, is to serve as a source of knowledge to be presented to the student. Its second function is to provide a standard for evaluating the students performance. It must be able to generate solutions to problems in the same context as the student so that answers can be compared. The module must also be able to detect any gaps in the students knowledge and be able to assess the students overall progress.

   One problem with this module and the manner in which it represents knowledge is that the designers view of the material is very much imposed on the user. This is true with most tutoring systems. It is the implicit representation of knowledge which distinguishes this method from the others.

2. The *student model module* refers to the dynamic representation of the emerging knowledge and skill of the student. If the tutoring system is to be considered intelligent then a profile of each student must be maintained. This is why ITSs are considered to be closer to what is required in a tutoring system. Human tutors maintain a profile of each of their students. They know their rate and style of learning so it seems obvious that the computer tutor should know these facts also. This model should include all those aspects of the students behaviour and knowledge that have a possible impact on their learning and performance. Building such a student model is virtually impossible however, as the only means of obtaining information on the student is via their performance throughout the tutorial. This is very restrictive when one considers the various ways available to a human tutor of assessing his/her student's profile, i.e. voice, facial gestures, etc. Difficult as it may be to construct such a student model, it is a worthwhile exercise as they perform several different functions :

   1. Corrective : eliminate errors in the student's knowledge
   2. Elaborative : correct incomplete student knowledge
   3. Strategic : initiate important changes in the tutorial strategy

4. Diagnostic : diagnose errors in the student's knowledge
5. Predictive : determine the student's likely response to tutorial actions
6. Evaluative : assess the student

The student model can also be used to assess how much the students have learned from using the tutoring system. The student model module can be compared to the expert knowledge module and if some gaps in the students knowledge were noted, instruction could be geared towards these areas. This module thus serves a dual function : that of acting as a source of information about the student and serving as a representation of the student's knowledge.

3. The *tutoring module* is the part of the ITS that designs and regulates instructional interactions with the student. It is closely linked to the student model, using knowledge contained within this module to help it to determine which teaching strategy to apply to each individual student. This module does not really require any domain expertise built into it. It is concerned with what responses the student generates and whether these responses match what the expert knowledge module would generate, not what these responses mean. The tutoring module tailors it strategy to the needs and learning styles of each individual. Some students prefer to be guided through the knowledge whereas others prefer to roam around the network and discover the knowledge for themselves. The tutoring module can adapt to these differing requirements. This module is thus an integral part of the ITS. It provides the bridge between the experts knowledge and the student model. The manner of presenting the material contained in the domain knowledge module to the student can very much determine whether or not the information will be understood and retained by the student.

4. The *user interface module* is the communicating component of the ITS which controls interaction between the student and the system. This module translates the internal representation of the expert knowledge into a form which is understandable to the student. The user interface is a vital component of any ITS because if the student is not satisfied with the manner in which the information is presented then the system will remain unused, regardless of how "intelligently" the system chooses the material to be displayed. This is why so much product development time is invested in the interface component of many computer based training courses. With the advent of multi-media systems the designers job is made even more difficult because he/she has to decide not only what information to display but also what format it will take, e.g. text, video, graphics, etc. On the reverse side, the student finds it easier to fix his/her attention on a tutorial which is interspersed with video, sound and so on, instead of reading screen after screen of text. It is important that the interface be easy to learn and use as the student does not want the task of dealing with the interface in order to learn the subject matter of the expert knowledge module [ANDE90]. An easy interface is one that minimises the number of things to be learned and minimises the number of actions (e.g. mouseclicks, keystrokes) that the student has to perform in order to communicate with the tutor. The interface should also be as consistent as possible so that, with use, the student becomes very familiar with the actions required to manipulate the system.

ITSs seem to provide us with the solution to our original problem of how to supply each student with a tutorial equipped to meet their individual cognitive styles. Theoretically, ITSs should model both the knowledge and the student and provide a link between them in the form of a teaching strategy. However, we have seen the problems inherent in representing expert knowledge and compiling a complete user profile. The immediacy of feedback is an advantage of ITSs in that the closer feedback on an error is to the actual error then the more effective is the tutor as the student knows more readily the stages which led him/her to make the mistake in the first place. It also helps the student to get back on the right track as quickly as possible thus avoiding the possibility of their forming erroneous ideas and concepts. This feedback should be carefully designed so that the student is forced to work out the correct answer rather than just see it on screen. A tutoring system must incorporate all three disciplines of computer science, psychology and education in equal measures before it can be deemed to be on a par with its human equivalent.

4. Learning from Hypertext

As mentioned earlier there have been many attempts at using hypertext techniques to organise information for presentation to users for learning purposes. As we have seen in an earlier section of this paper, learning as a process is highly contextualised and one could speculate that hypertext could therefore be a viable computer based learning medium as one of its richnesses is the fact that it contextualises information by describing a concept as a node and the interactions and relationships between that concept and related concepts as information links. In hypertext this inter-concept linking is done much better than it could ever be done in a linear form of the same information.

Current hypertext systems encourage hypertext authors to design and structure their hypertexts in a non-strict hierarchy. Such a hypertext would consist of some introduction node linked to some overview nodes, linked in turn to more detailed descriptions, eventually linked to leaf nodes on the tree which would contain detailed information and which would be linked to other "leaf nodes". A schema for such an architecture is given in Figure 1. The non-leaf or overview nodes in this organisation would correspond to a table of contents in a paper-based or linear document.

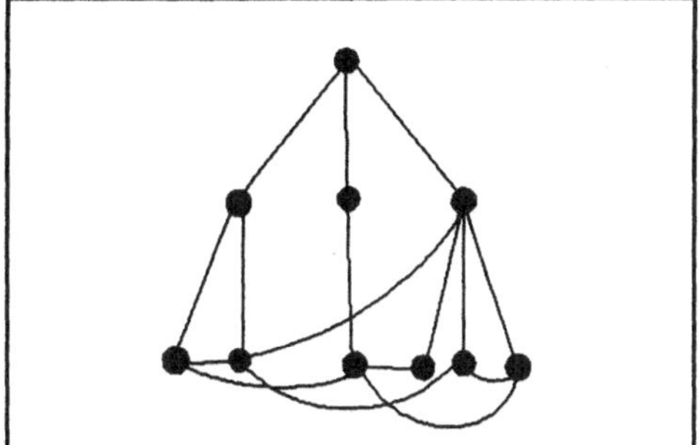

Figure 1: Typical hypertext organisation

In designing our hypertext for use as adjunct courseware in an undergraduate course on databases, we took a slightly different approach. Basically our information organisation was very similar in that we also had an overview node, but our topic overviews were not organised hierarchically, but as a network, as shown in the schema in Figure 2. Our hypertext consisted of 462 nodes (almost 250k bytes of text) and 1398 one-way information links, making it a small to medium sized hypertext.

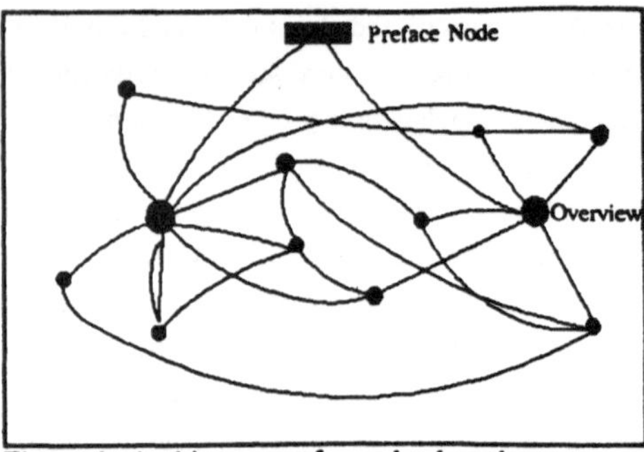

Figure 2: Architecture of our database hypertext

There is a very real danger of designing hypertexts specifically to operate as teaching/learning tools and in doing that, constructing node contents which encourage users to follow specific paths or trails through the information space. Such a system then becomes structured CBT, without the testing of users, What we want to find out is how hypertexts could support the learning process. Incidental learning of material from a hypertext would be equivalent to browsing, but can hypertext support discovery learning through its representation of the context of information ?

Our experiences in creating and using our hypertext for databases have shown us many things. In particular they have illustrated the inadequacies of the current generation of hypertext products, especially when it comes to creating hypertext material for learning purposes. Our survey of the use of the hypertext has shown that it has been used more for reference than for learning, which is not surprising really [SMEA91]. Our hypertext was not really useful for deep understanding as to acquire deep understanding would require student participation in some DBMS-related processes. The hypertext we used did not provide any user participation; it is a passive presentation of information without getting the user involved in answering questions, etc. Conventional CBT techniques would be better than current hypertext products in promoting deep learning.

Some attempts are being made to extend hypertext in the direction of supporting the learning process. The SEPIA project [STRE90] focusses on the production of argumentative texts within the hypertext information space. The contention is that when designing material for learning, using hypertext, CBT or paper-based, the author wants to convince the student of something. By presenting argumentative texts the SEPIA project is supposing that they are good for learning beyond the introductary level. Stretiz *et al.* believe that acquiring a deep understanding of concepts and reasoning about them does require active involvement of the learner and conjecture that argumentative texts in hypertext form present an opportunity to get a user actively involved in the topic by allowing him to make his own arguments with personal annotations to the hypertext and exploitation of personal information linking. This isn't a cognitive theory of learning, it is simply an attempt to exploit the known fact that deep learning requires student participation and that hypertext systems are restrictive in allowing user participation.

What all this means for current hypertext systems as vehicles for learning is that they can provide contextualisation better than could linear material or conventional CBT. Hypertext is worse for deep learning than CBT or conventional courseware as it cannot enforce discipline and hypertext tools are completely passive and inactive systems. Relying on the user/learner to do the learning from the hypertext requires a user's prior knowledge, learning and thinking, which is not always available but is acquired in intelligent tutoring systems. A second approach to addressing this would be to develop a delivery system which is intelligent in that it knows about and tries to realise learning goals. To do this it should:

1. Be able to diagnose users' prior knowledge
2. Be able to diagnose users' operating skills
3. Be able to diagnose users' state of comprehension
4. Be able to diagnose users' misconceptions
5. Be able to model the user.

Hypertext systems fall way short of this target. Hypertext systems cannot accomodate individual user models as do tutoring systems so the bounds on what is achievable by using hypertext for learning are quite obvious.

The SEPIA project, like almost all work on hypertext for learning, does not enforce any discipline on students. But hypertext, by its very nature, encourages user browsing and freedom of choice but we need to enforce some control on users if they are to become students who learn from the hypertext. Conventional CBT enforces discipline by insisting on students attaining certain scores in tests before they can make progress. Hypertext browsing should not, nor could not, support user testing in this way. In our work we believe we have found a technique for allowing structured, disciplined browsing in an unstructured hypertext, which would be conducive to supporting learning from hypertext. By dynamically creating guided tours for individual students we can provide discipline on students, ease the problems of disorientation in browsing, and cater for individual student needs by tailoring guided tours to individuals. Furthermore, by dynamically amending tours during browsing we can also cater for shifting information needs. We shall describe our approach in the next section.

5. Guided Tours for Learning

Mention has been made in this paper of the idea of guided tours in a hypertext and static guided tours have been offered as a solution to the disorientation problem as discussed earlier. In some hypertext systems a hypertext author has the facility to create a recommended path through the hypertext, in addition to the normal information links. The tour is what the author of the hypertext feels would make a good series of nodes to view in a certain order. A reader can follow the authors recommended tour if she desires, or can follow other information links if that preference arises. As a solution to disorientation, a user can always return to the guided tour if she gets lost, so there should be no inhibition in browsing around the hypertext. Thus a guided tour can be thought of as a kind of "superlink" which connects a string or chain of nodes instead of just two nodes [NIEL90].

Guided tours are not only useful in solving the disorientation problem but they may also have a role to play when hypertext is being used as a teaching aid. Allowing a student to browse around an information space would have advantages but the purpose of a tutorial is to impart information and so a scheme should be implemented whereby the student cannot end the session unless certain important nodes have been visited. One way of doing this would be to incorporate some AI techniques into the system which can determine if the relevant nodes have been studied during a browse or not. Another approach would be to use guided tours. If the system made it compulsory for the student to view all nodes in a tour and answer questions at the end as with conventional CBT, then the author of the hypertext system would be sure that at least the students were studying the important facts. However this would undermine the freedom of choice of hypertext use if the user were made follow the tour all of the time.

Two specific examples of using guided tours for disorientation will illustrate how they work. One of the most well known implementations of the guided tour strategy is that developed by Xerox PARC in the NoteCards system [TRIG88]. NoteCards uses *tabletop cards* or screen dumps as building blocks for the guided tour. The tabletop card structure is used to capture the layout of the screen at a particular time. The guided tour is a graph whose nodes are tabletop cards and whose links are special guided tour links connecting these cards. The author of a new guided tour creates new tabletop cards and links them together to form the paths making up the tour. Readers of hypertexts wishing to use the tour facility can begin by clicking on a series of buttons. START causes the first tabletop in the tour to be displayed, NEXT moves along the tour to the next tabletop card, PREVIOUS provides a history list of all the tabletop cards in the tour already visited, etc.

PERSEUS is a hypertext system designed to teach Greek history and literature to University students [MYLO90]. It utilises a guided tour facility to store a pre-determined set of locations which can be visited again and again. An annotation facility is added to the tour which allows readers to save locations of interest and also to comment on the contents of these locations. A reader can join a guided tour from within any node in Perseus and when the option to join a tour is selected, a list of tours available is displayed and the reader must select one. While following a tour, the annotation made by previous readers are also available.

So far in this paper we have discussed the use of guided tours both as a means of solving the disorientation problem and as a means of ensuring that important sections of a hypertext tutorial are covered. In our work at Dublin City University we are trying to devise intelligent methods of searching through a hypertext and we believe that implementing a **dynamic** guided tour facility which would automatically choose the best **route** rather than just a starting node for a user to take, based on a user's search query, is a promising avenue to pursue. This idea differs from methods previously discussed in that the route taken through the hypertext is computed only when the user enters a search query or expression of information need, whereas the static tour consists of author-defined paths. The dynamic tour has the advantage of being directly related to the needs of the user where only relevant nodes are displayed as opposed to the static case where the author decides what is relevant before the users have even formulated their queries. Both computer based training and hypertext technology are interleaved in this strategy and the dynamic approach to creating a guided tour offers the user the best of both worlds in that they can follow the guided tour and if they wish to navigate the hypertext on their own they can leave the tour at any time.

In order for a dynamic tour to be created, the user has to enter a natural language search query. This query is then parsed, stopwords are removed and the remaining words are stemmed. A method of finding the most suitable nodes to add to the guided tour will have to take into consideration the frequency of the stemmed query terms in each node, how well connected is each node, the contents of the nodes neighbours, etc. Because the guided tour is to be used in a training environment, the order of presentation of hypertext nodes, i.e. the overall route, is very important. There is no point in selecting the most relevant nodes and displaying them in a random order if the user is trying to learn about the subject matter as well as just retrieving information. There should be a natural progression from the most basic information to the more complex so that the user does not become confused with the material being presented. The usefulness of dynamic guided tours can be seen when one considers what would happen in a training environment where the student is trying to teach himself with no teacher or supervisor present. If the student merely navigates through the hypertext information with no set structure on his browsing, the material becomes just a blur of information which the student finds difficult to retain, referred to as the *Art Museum Problem* in hypertext literature. A dynamic guided tour offers the student the choice of logically progressing through the material in a manner designed by the system to best make use of the information available and this is done uniquely for each user's search. When the author of the hypertext is not available (as is usual) and there is no tour facility, the user is left to himself to muddle through the network as best he can.

In attempting to improve the effectiveness of the dynamically created guided tours in terms of satisfying the user, the system should assess the opinions of the users on the guided tour and take these into account when planning the route. This feedback operation could take place while the user is following the tour and, depending on whether they consider the nodes in the planned route to be relevant or not, the route could be changed automatically to fall in line with the changing or perhaps the more specific user needs. The user could view the information relating to the already retrieved nodes such as titles or abstracts and use these details to modify their original query statement, normally by adding terms or phrases from nodes that appear relevant and by deleting terms and phrases included in nodes considered irrelevant [SALT83]. If this relevance feedback operation is carried out several times in the course of following a tour, the set of nodes which are retrieved each time more reasonably approximates the set of relevant nodes in the entire hypertext.

Relevance feedback has been used quite frequently in conventional information retrieval systems with consistent success. However, applying this strategy to a hypertext system may prove to be more difficult because of the rather unstructured layout of the nodes. Retrieval of a document is no longer a straight-forward matter of finding the node with the highest number of terms in common with the query. With hypertext we have to consider items such as number of connections, relevance of neighbours, etc. This could make the implementation of a relevance feedback mechanism very complex but work in this direction, though only for computing a start node, has been done by Frisse as discussed earler [FRIS88].

One of the dangers of simply applying information retrieval techniques (stemming, relevance feedback, etc) to hypertext tour creation is that the resulting access method is information retrieval from hypertext. In order to add some "intelligence" to the process we propose to introduce some planning elements into the tour construction module. A plan is a program of action or sequence of steps, designed to achieve a particular goal [GENE82]. Plans have an important part to play in the context of intelligent tutoring systems as we have already seen. The overall objective of any computer based tutorial is to impart knowledge to the user, not merely to present information. At the end of a tutorial session the student should have learned something about the presented material and an intelligent tutoring system would lead the user through the information in a logical manner designed to help the student retain all the material based on a student and/or information model.

The dynamic creation of guided tours should follow this type of logical progression so that the user is not jumping from one document to another with no obvious connection between them and to ensure that a particular sequence is adhered to when necessary. For example, if a student is learning about local area networks, it wouldn't be sensible to have as the starting point in the guided tour a document on inter-networking, then LAN protocols such as CSMA and Token Ring and finally an introductory node on LAN's. If a student is only learning about such a topic, it is logical that the tour present them with an introduction to the area first, then discuss some protocols and finally, address more complex issues such as internetworking. This is where node context identification comes into play, as previously discussed. Not only must a node be related to a users query before it is selected for inclusion in the guided tour but its position in the tour should depend on its relation to the other selected nodes. The ultimate advantage of such a system would be that the overall discourse of the material presented in the tour would not appear as disjointed or fragmented but would have an general coherence.

Another aspect of intelligent planning which we can develop is for the hypertext system to tailor its information presentation to the needs of the individual student. No intelligent tutoring system can operate without an understanding of the student and so a student model should be formulated which would represent various characteristics of the student's behaviour and learning patterns [NWAN90]. This enables the ITS to identify certain areas with which the student is experiencing difficulty and also areas which the student finds easy to learn. The dynamic guided tour method of searching through hypertexts could use relevance feedback from users to modify its planned tour accordingly. In order for hypertext to be used effectively in a learning environment, note should be taken of context as well as content of hypertext nodes and a user profile should be maintained so that each individual can adapt the tutoring system to their own special needs.

6. Conclusion

Hypertext is a method for organising information which has recently attracted a lot of attention and has been applied to a number of diverse applications, including learning. In this article we have looked at learning and we have looked at hypertext and we have identified how hypertext could have a role in using computers for learning *if the functionality of current hypertext systems is extended.*

We have some specific proposals for amending what hypertext can do and we have articulated these in this paper. In order to make progress with our ideas we are working with the Telecom Eireann Training Division to develop a prototype system which constructs such dynamic guided tours. The hypertext being built covers training material on equipment like digital exchanges and PBXs. The underlying hypertext system is HyperDoc which runs on PCs or PS/2s. We are extending the functionality of the underlying hypertext system by allowing the typing of nodes and of links, in order that the planner that we are developing can use this meta-information to judiciously construct routes. We expect to have the first prototype completed by the end of this year (1991).

The prospects for using hypertext in computer assisted learning are quite good. In the short term our work should improve the contextualisation in hypertext via guided tours. In the medium term hypertext learning systems should include the use of outside tools, within hypertext nodes, to provide user participation leading to deep learning. An example of this would be a hypertext on databases where a student views the contents of a node and is actively involved in inputting some SQL which is executed dynamically on an underlying relational database to provide the answer to the SQL query. In the long term there should be a cognitive model of learning and of

authoring which should lead to better and more well-founded hypertext design guidelines. It should be noted though that our present aspirations of using hypertext for learning are quite modest in terms of what can be done with sophisticated intelligent tutoring and that dynamic guided tours are just one step in the direction of using hypertext information for learning in an intelligent manner.

Bibliography

[ANDE90] Anderson J., Boyle C., Corbett A.,et al, Cognitive Modeling and Intelligent Tutoring, Artificial Intelligence, Vol. 42, 1990, pp 7 - 49.

[BRAT88] Bratko I., Lavrac N.(Eds), Progress in Machine Learning, Sigma Press, 1987.

[DEAN83] Dean C., Whitlock Q., A Handbook of Computer Based Training, Kogan Page, London, 1983.

[FRIS88] Frisse M., Searching for Information in a Hypertext Medical Handbook, Communications of the ACM, Vol. 31, No. 7, July 1988, pp. 880 - 886.

[GAIN88] Gaines B., Boose J., (Eds), Knowledge Acquisition for Knowledge Based Systems, Vol. 1, Academic Press.

[GARG90] Garg P., Scacchi W., A Hypertext System to manage Software Life-Cycle Documents, IEEE Software, May 1990, pp. 90 - 98.

[GENE82] Genesereth M., The role of plans in Intelligent Teaching Systems, In: Sleeman, D. & Brown, J.(eds), Intelligent Tutoring Systems, Academic Press, 1982.

[HALA88] Halasz F., Reflections on NoteCards : Seven Issues for the Next Generation of Hypermedia Systems, Communications of the ACM, Vol. 31, No. 7, July 1988, pp. 836 - 852.

[LAVE88] Lave J., Cognition in Practice: Mind, Mathematics and Culture in Everyday Life, Cambridge University Press.

[MYLO90] Mylonas E., Heath S., Hypertext from the data point of view : Paths and Links in the Perseus Project, In: Rizk, A., Streitz, N. & Andre, J. (eds), Hypertext : Concepts, Systems and Applications, pp 324 - 336, Cambridge University Press, 1990.

[NIEL90] Nielson J., Hypertext and Hypermedia, Academic Press, London, 1990.

[NWAN90] Nwana H., Intelligent Tutoring Systems : an Overview, Artificial Intelligence Review, (1990) 4, pp. 251 - 277.

[RAYM88] Raymond D., Tompa F., Hypertext and the Oxford English Dictionary, Communications of the ACM, Vol. 31, No. 7, July 1988, pp. 871 - 878.

[SALT83] Salton G., McGill M., Introduction to Modern Information Retrieval, McGraw Hill Computer Science Series, 1983.

[SMEA90] Smeaton A., Experiences in Creating a Hypertext for a Database Course, School of Computer Applications, D.C.U., Working Paper : CA-19-90.

[SMEA91] Smeaton A., Using Hypertext for Computer Based Training: Results and Analysis, Computers in Education, (submitted).

[STRE90] Stretiz N., Hannemann J., Elaborating Arguments: Writing, Learning and Reasoning in a Hypertext Based Environment for Authoring, In: Designing Hypermedia for Learning, D.H. Jonassen and H. Mandl (Eds), NATO ASI Series F67, Springer Verlag.

[TRIG88] Trigg R., Guided Tours and Tabletops : Tools for Communicating in a Hypertext Environment, ACM Transactions on Office Information Systems, Vol. 6, No. 4, October 1988, pp. 398 - 414.

PELICAN: A Prototype Information Retrieval System using Distributed Propositional Representations.

Richard F.E. Sutcliffe[1]

Department of Computer Science

University of Limerick, Ireland

Abstract

A long standing issue within Natural Language research is how the meaning underlying language can be represented in a computationally tractable form. A great deal of work has been based directly or indirectly on the case structures originally proposed by C.J. Fillmore. While Fillmore's ideas have been shown to have many strengths, a number of significant problems arise when they are implemented within a conventional symbolic processing paradigm. In this paper we describe an alternative approach where a case structure can be converted into a single distributed pattern. This is accomplished by the use of distributed representations for both semantic cases and fillers. The representations have been incorporated into an experimental Information Retrieval system called PELICAN which can convert an information need about UNIX expressed in English into an ordered list of help files which the user can consult in order to obtain the information he or she requires.

1 Introduction

Within Natural Language Processing (NLP) research a considerable amount of attention has been focussed upon representational schemes for capturing linguistic meanings in a fashion which is tractable computationally (eg Bruce 1975 [1]). The reasons for this are easy to see. When processing language we are primarily concerned with what is meant and only secondarily are we interested in exactly how that information is conveyed. Similar meanings can be conveyed in a myriad of different ways in any natural language. Moreover speakers of a language have an intuitive notion of semantic similarity which they can apply to any pair of sentences in order to judge how similar in meaning they are. It is natural within NLP therefore, to try to convert sentences which one wishes to process into structures which reflect underlying meaning directly. That way, queries can be processed, inputs matched against knowledge structures or whatever else is required in order to process the input, without having to consider the manner in which the original meanings were communicated.

[1] I am grateful to Paul Mc Kevitt for reading an earlier version of this paper and I would like to thank John Gooday and Myles Mylvaganam for their help with this research. I am also indebted to Khalid Sattar for providing excellent system support at Exeter, and to Tony Molloy who has gone too enourmous trouble to set systems up for me at Limerick.

It is convenient to divide variations in the expression of the same meaning into two broad classes. The first class is concerned with amorphous concepts expressed by single words and noun phrases. There are often many different words in a language for conveying similar concepts, eg 'dog', 'hound', 'wolf', 'beast', and so on. In addition there may be a large number of noun phrases for expressing the same idea, such as 'man's best friend', 'canine animal', 'domestic quadruped', etc. The problem for NLP is to convert all such words and phrases into a similar structure which, in this case, captures the dog concept.

The second class of variations exists at the syntactic level. Even given the same choice of words for the fundamental concepts, many different structural variants may be employed in expressing an idea. One well known example is passivisation, eg 'The dog bit the cat' vs. 'The cat was bitten by the dog'. Another example is an assertion of the same event, such as 'Truly it is the case that the dog bit the cat'. All three sentences capture the same idea, but how can this be expressed?

One of the best known approaches to the problem of amorphous concept representation has been to use a set of *features* of Katz and Fodor (1963) [2], Chomsky (1965) [3], Rosch (1973) [4] or *semantic primitives* (see Wilks (1975) [5], Lehnert and Burstein (1979) [6]. However, there are many difficulties. For example, how should a set of features be chosen, how can concepts be encoded in terms of those features and how can such concepts then be compared? In particular, should the features take part in a compositional calculus of some sort (eg the Conceptual Dependency of Schank (1972) [7]) or the Preference Semantics of Wilks (1975) [5] or not?

One of the best known attempts to tackle the problem of representing structural variants of sentences in a canonical fashion is the work of Fillmore (1968) [8]. Fillmore proposed that in addition to specifying the action underlying a sentence, other pieces of information regarding it could be categorised under a small set of predefined headings called *semantic cases*. For example in the sentence 'John hit Bill with an umbrella' we might classify John as the *Agent*, Bill as the *Patient* and the umbrella as the *Instrument*. The hypothesis underlying this work was that the concept underlying each syntactic constituent within a sentence (ie noun phrase, prepositional phrase and so on) would be assigned to a unique case. Thus a pair of sentences which conveyed the same meaning would be transformed into identical case frames regardless of any syntactic differences.

One of the best known implementations of Fillmore's work is the Conceptual Graphs of Sowa (1984) [9]. The Conceptual Dependency of Schank (1972) [7] and the Preference Semantics of Wilks [5] are also based on very similar ideas.

The approach taken in our work is also to use semantic case frames for representing sentential meanings. However, rather than using a list-based representational scheme as with the work of Sowa, Schank and Wilks, we use distributed representations, both for capturing amorphous concepts and for case structures. There are a number of reasons for doing this. Firstly, the use of distributed patterns for concept representations allows us to capture some of the subtle and fluid aspects of language which elude those who use other approaches. Using distributed representations we can capture an arbitrary number of concepts and we can also handle indefinite ideas like 'charity' or 'friendship' alongside concrete objects such as 'bicycle'. This is hard to do in other approaches. In addition we can compare meanings using simple arithmetic techniques, as we shall show later.

Secondly, by using distributed representations for case structures as well as concepts we can get around a major problem with Fillmorian schemes: in practice there is considerable ambiguity as to which semantic case should be used to capture a given syntactic constituent. For example in the sentence 'Susan gave Mary a book' is Susan the *Source* or the *Agent* and is Mary the *Beneficiary*, *Patient* or *Destination*? By using distributed representations we effectively have a whole spectrum of semantic cases (rather than a small set) to which we can assign our constituents without losing the vital property of similar case structures for similar meanings.

In this paper we present an experimental NLP system, PELICAN, which uses distributed case frame representations. The system performs a simple Information Retrieval (IR) task relating to the UNIX system. A query such as 'I want to move some files' is input to the system in English. The system then outputs a list of UNIX help files ordered by their relevance to the query.

A crucial advantage of using distributed patterns for capturing meanings is that the semantic content of a pair of sentences are very easily compared by a simple operation performed on their corresponding representations. This feature is central to PELICAN's operation.

The work relates closely to two areas of IR research. In the well known Vector Space Model of IR (Salton 1971 [10]) both queries and documents are represented as vectors and the retrieval operation comprises the comparison of the query vector with each document vector. PELICAN is using a similar technique, the difference being that our vectors are capable of capturing full semantic case frames rather than just weighted terms.

A number of recent IR projects have experimented with sophisticated representational schemes for capturing the meanings of both texts and queries (Croft 1987 [11]). Examples are the Tree Structured Analytics of Smeaton's SIMPR project (Smeaton 1989 [12], Smeaton and Sheridan 1990 [13] and Smeaton, Voutilainen and Sheridan 1990 [14]), the REST structures in ADRENAL (Lewis et al 1989 [15]), and the structures in the TOPIC system (Hahn and Reimer 1986 [16], Thiel and Hammwohner 1987 [17]). The aims of these projects are very similar to our own. However one of the disadvantages of these approaches is that pattern matching techniques must be developed to compare queries with documents. In our scheme, the matching is very simple and falls out of the representation.

In the rest of the paper we present in outline a summary of what has been accomplished in the PELICAN project so far. In Section Two we start off by summarising the techniques we are using for capturing amorphous concept representations. Then in Section Three we turn to case frames and show how they can also be captured using distributed representations. Section Four describes PELICAN and explains how the system uses these distributed representations. Preliminary results for the project are also presented. Finally, Section Five provides a summary, discusses our conclusions and outlines some possible next steps for the project.

2 Using Distributed Representations for Concepts

In this section we briefly outline the method used to represent the meaning of amorphous concepts in PELICAN. These representations will form the fillers in our distributed case frame representation. Our strategy is to use a fixed set of features each of which expresses an attribute which an object might have. For example suppose for simplicity we consider a set of three features, *is_furry*, *has_wings* and *is_intelligent*, and we wish to express the concept 'person'. This is easily achieved by stating which of our three features is possessed by a person:

person	
is_furry	no
has_wings	no
is_intelligent	yes

In many cases the application of an attribute to a concept is not clear cut — an attribute may for example apply more strongly to one concept than to another. We can capture this idea within the same representational scheme by associating a numeric *centrality* with each feature, rather than a simple yes or no, which expresses the extent to which the attribute captured by the feature applies to the concept. In our example we might encode 'person' like this:

person	
is_furry	1
has_wings	0
is_intelligent	10

Now consider how two other concepts, 'cat' and 'bird' can be encoded using the same three features:

cat	
is_furry	10
has_wings	0
is_intelligent	5

bird	
is_furry	4
has_wings	10
is_intelligent	3

Since each concept is encoded in terms of the same set of features we can consider each representation as a vector in a three dimensional space. If each vector is *normalised* (ie scaled up or down so that its length is one) then it will effectively describe a point on a sphere. This is shown in the diagram. The interesting point is that vectors corresponding to concepts which are similar in meaning will lie close together on the sphere while more disparate concepts will be further apart. Thus if we were to encode the concept 'individual' we would expect it to lie close to 'person' on the sphere, as shown in the diagram.

The advantage of this approach is that similarity of a pair of meanings is directly correlated with similarity of a pair of vectors. Thus in order to compare meanings we can simply use any vector comparison method. In this work we use the dot product.

Naturally, a system involving only three features has rather limited ex-

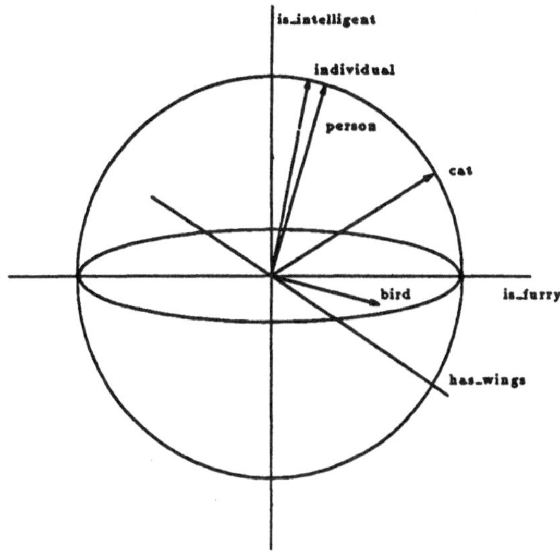

Amorphous concepts mapped onto a sphere

pressive power. However the same ideas generalise to a system of n features. In PELICAN we use two sets of features, one for representing actions (as described by verbs in English) and the other for capturing objects (as described by nouns and noun phrases). These feature sets were developed for an earlier system called PARROT which performed a primitive summarisation task in completely different domain — stereotypical domestic events (Sutcliffe 1988a, 1988b [18,19]).

The use of feature sets in constructing distributed concept representations is of course not new. One of the earliest discussions is that of Waltz and Pollack (1985) [20], while similar ideas have been taken up elsewhere (eg Rumelhart and McClelland 1986 [21]). There are of course many issues relating to distributed concept representations — for example, how many features there should be, how concepts should be encoded, the relationship between labelled and unlabelled feature schemes, and so on. These are addressed in more detail in Sutcliffe (1991a) and (1991b) [22,23]. There is also an extended treatment in Sutcliffe (1988a) [18].

There is a strong similarity between what is being discussed here and the Vector Space Model (VSM) of Salton (1971) [10]. In the latter case each element in the eponymous vector is a *term* rather than a feature, and the objective is to retrieve documents whose vectors are similar to a query vector. In fact the PELICAN system operates on very similar lines. The difference is that we can capture full case structures with our vectors. In the next section we describe how this is done.

Case Feature	Filler Feature
causer	object features
doer	object features
done_to	object features
experiencer	object features
source	object features
destination	object features
instrument	object features
place	object features
action	action features

A table showing the semantic case features used in PELICAN. The type of filler features corresponding to the various cases are also shown.

3 Distributed Case Structures

In this section we turn to the implementation of a Fillmorian representation scheme using distributed representations. Suppose we wish to encode a sentence such as 'Gerry went to Tipperary by car'. A typical list based approach to the problem would be to enclose the entire frame in a list and to pair semantic cases together with their fillers:

((Action go) (Agent Gerry) (Destination Tipperary) (Instrument Train))

Our approach is analogous to this idea, the main difference being that we use distributed patterns for both semantic cases and their fillers. Filler patterns are defined either as vectors over the set of action features or as vectors over the set of object features, depending on the case which we are filling. Semantic case patterns are defined analogously over a set of semantic case features. The set used in PELICAN is shown in the figure.

The objective in using distributed representations for semantic cases is to allow a given filler to be assigned partially to a number of different cases. Thus to assign unambiguously to a single case we do not need a distributed pattern, only a localist one. Consider the problem of assigning 'Tipperary' to the destination case for example. Using the case features shown in the diagram, the correct case pattern will simply be [0,0,0,0, 0,1,0, 0,0]. However, consider the example given earlier of 'Susan gave Mary a book'. We can make Susan have both Source and Agent case (**Source** and **Doer** in our system) by means of a pattern such as [0,0.707,0,0, 0.707,0,0, 0,0]. The relative weighting between the various cases can of course be controlled as we see fit.

Having considered how to generate ⟨case, filler⟩ pairs where both the case and filler are distributed patterns, the next task is to group them together. In this work we follow Smolensky (1987a, Dolan and Smolensky 1989) [24, 25] in

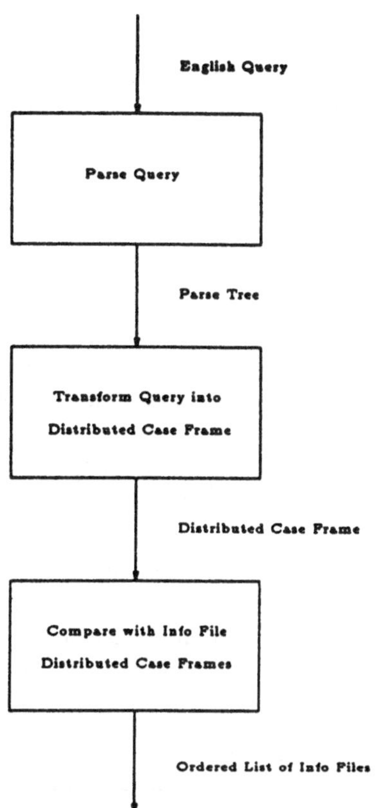

Architecture of the PELICAN system

```
[sentence, [noun_phrase, [opt_pronoun, [pronoun, i]], [], [], [], []],
[verb_phrase, [verb_group, [], [], [], [verb, want]], [noun_phrase_group,
[noun_phrase, [], [], [], [], []], [noun_phrase, [], [], [], [], []]],
[verb_phrase, [verb_group, [opt_modal, [modal, to]], [], [], [verb, move]],
[noun_phrase_group, [noun_phrase, [], [opt_determiner, [determiner, a]],
[], [], [noun_star, [noun, file], []]], [noun_phrase, [], [], [], [], []]],
[prepositional_phrase_star, [prepositional_phrase, [preposition_plus,
[preposition, from], []], [noun_phrase, [], [], [], [adjective_star,
[adjective, one], []], [noun_star, [noun, directory], []]]],
[prepositional_phrase_star, [prepositional_phrase, [preposition_plus,
[preposition, to], []], [noun_phrase, [], [], [], [adjective_star,
[adjective, another], []], [noun_star, [noun, place], []]]], []]]]],
[sentence_terminator, .]]
```

Parse tree for query 'I want to move a file from one directory to another place'.

using the outer product operation. Thus if we have a case **c** and a filler **f** they are simply combined to form a ⟨case, filler⟩ pair $\mathbf{c}\mathbf{f}^T$. Note that both vectors are normalised (scaled so that their length is one) before performing this operation.

Having produced a set of case filler pairs $\mathbf{c}_i\mathbf{f}_i^T$ each of which is a matrix, we can construct a representation for the entire proposition **P** simply by adding these matrices together:

$$\mathbf{P} = \sum_i \mathbf{c}_i\mathbf{f}_i^T$$

The advantage of using distributed case structures of this kind is that a pair of such structures can easily be compared. Thus we can determine how similar in meaning two sentences are by comparing their corresponding case structure patterns. The dot product can be used for this purpose by using it to compare the filler of each role in the first proposition (case structure) with its corresponding role in the second proposition. It is convenient to scale the entire proposition after performing the summation above and before performing this comparison, so that the result of comparing a proposition with itself in this way is one. Having looked briefly at the propositional representation which we are using, the next step is to consider how such structures are used within the PELICAN system.

4 The PELICAN System

4.1 Outline of Operation

A major objective of the PELICAN project is to experiment with propositional structures. The domain of the project is Information Retrieval. PELICAN aims to help people decide which UNIX help files they should consult in order to satisfy a particular information need. This need is expressed as a simple query in English stating what the user aims to do, for example 'I would like to send mail to someone'. In response to the query the system produces a list of help files which are ordered by their relevance to the query. It is assumed that the user will search for enlightenment in the first help file on the list. If the answer is not found there, the user will consult the second file, and so on. The operation of PELICAN is very similar in principle to the Vector Space Model which has been widely researched within IR (eg Salton 1971 [10]). Corresponding to each help file which the system is apprised of, there exists a pre-encoded proposition of exactly the type discussed in the previous section. This proposition captures the gist of the information contained in the help file. A query input by the user is also converted into a proposition. The retrieval operation simply involves comparing the proposition corresponding to the input query with each of those corresponding to information files and subsequently ordering the files by their degree of match. The assumption behind this work is therefore that one can match the 'gist' of a query with the 'gist' of an information file. The figure shows the architecture of the present system which operates on the following lines.

Firstly, the input query is converted into a symbolic parse tree by a standard Definite Clause Grammar (Pereira and Warren 1980 [26]). Next the symbolic parse tree is transformed into a symbolic weighted case structure from which a distributed proposition can be generated. In the final stage, this proposition is compared with the propositions corresponding to help files. Consider for example the query 'I want to move a file from one directory to another place'. The symbolic parse tree produced by the system for this query is shown in the figure and the symbolic weighted case structure into which it is transformed is shown in the following figure. The weighted case structure consists of a series of ⟨case, filler⟩ pairs. Each semantic case is expressed explicitly as a pattern over the set of case features. At present each case pattern derived from a parse tree is localist, that is it involves only one of the possible cases. However this is not the case for the representations of information files. The semantic case pattern is converted by the system into a normalised vector over the set of semantic case features. Each filler is expressed as a word in the lexicon where it is pre-defined as a pattern over a suitable set of features.

The next figure shows a typical representation of an information file, in this case the one for the UNIX command 'ls'. As can be seen, this is in essentially the same format as the query. Note that the filler 'contents_of_directory' is being partially assigned to three cases, mainly to **done_to**, but also partially to **experiencer** and **place**.

4.2 Converting a Parse Tree to a Proposition

PELICAN is essentially a hybrid system because it uses both symbolic techniques, such as parsing based on grammar rules, and distributed representations. The two paradigms are quite different and therefore a major problem for such systems is to accomplish an effective interface between the two types of processing. In our case this problem is simplified by the fact that it is one way — we convert from a query in English to a proposition but never the other way round. This is because of the simplicity of the task which we are at present addressing. However, the task of converting from a symbolic parse tree to a distributed pattern is by no means straightforward, and what we present here is only a very simple first attempt at providing a solution.

At present we have chosen to the limit the task to a very restricted class

```
summary { a query }
case { action 5 }
filler move
case { done_to 5 }
filler file
case { source 5 }
filler directory
case { destination 5 }
filler place
```

List based weighted case structure for the query 'I want to move a file from one directory to another place'.

of parse trees. Effectively our queries at present are limited in syntactic form to simple declarative statements of a wish, the wish itself being expressed as an infinitive complement to the main verb. Thus in our example query 'I want to move a file from one directory to another place' we only need to work with the complement together with its associated prepositional phrases, ie 'to move a file from one directory to another place'. As will have been deduced from the preceding figures, the task of conversion to a pattern is divided into two stages. Firstly, we produce a weighted case structure which comprises a list of ⟨case, filler⟩ pairs. Secondly we convert this structure into a proposition by performing the simple vector operations which were discussed earlier. Each case specification is of course a distributed pattern over our set of semantic case features. At present, however, we have made a considerable simplification: a semantic case pattern which is derived from a parse tree is always *localist*, that is only one case feature within that pattern has a non-zero value associated with it. Therefore in the present situation, our problem of converting the parse tree to a proposition reduces to the assignment of syntactic constituents to individual cases as represented by our set of case features.

The syntactic constituents which are of interest in this endeavour are the verb from the complement, any noun phrases which follow that verb, and any prepositional phrases which follow that verb. The verb is used to fill the **action** case. We use the representation of the verb which is found in the lexicon as the filler for this case. Thus, continuing our example, we generate the structure `case { action 5 } filler move`. The integer five is arbitrary in a localist pattern as it will be converted to a one by the normalisation process. In our notation `filler move` means that the distributed pattern corresponding to the verb move in the lexicon is to be used as the filler for this case.

Turning to the noun phrases following the verb, at present our convention is to transform the first of these into a filler for the case **done_to**. Thus '...a file...' is transformed into `case { done_to 5 } filler file`.

Finally, we transform a prepositional phrase starting with 'from' into a **source** case and a prepositional phrase starting with 'to' into a **destination** case. Thus '...from one directory...' becomes `case { source 5 } filler directory` and '...to another place...' becomes `case { destination 5 } filler place`. As before, the integer centrality five is arbitrary since all these case patterns are localist.

An important point should be noted in relation to the translation of noun

```
command ls
summary { List contents of directory }
case { action 5 }
filler list
case { done_to 5 experiencer 1 place 1 }
filler contents_of_directory
```

An example case structure representing a document, in this case the help file for the command 'ls'.

```
I want to move a file from one directory to another place.

 1 mv         ( 4) 0.707107
 2 more       ( 3) 0.616366
 3 rm         ( 9) 0.529781
 4 cd         ( 2) 0.497723
 5 pwd        ( 7) 0.494732
 6 ls         ( 1) 0.465481
 7 mail       ( 6) 0.461548
 8 rmdir      (10) 0.456890
 9 rsh        ( 8) 0.444299
10 rmail      ( 5) 0.441480


I want to read my mail

 1 rmail      ( 5) 0.707107
 2 more       ( 3) 0.548026
 3 ls         ( 1) 0.511942
 4 cd         ( 2) 0.429793
 5 rsh        ( 8) 0.412470
 6 pwd        ( 7) 0.352758
 7 mail       ( 6) 0.350016
 8 rm         ( 9) 0.306724
 9 rmdir      (10) 0.306724
10 mv         ( 4) 0.260635


I would like to remove a directory.

 1 rmdir      (10) 0.990290
 2 rm         ( 9) 0.887207
 3 rsh        ( 8) 0.737982
 4 pwd        ( 7) 0.716869
 5 ls         ( 1) 0.686783
 6 mv         ( 4) 0.653809
 7 mail       ( 6) 0.650262
 8 more       ( 3) 0.629431
 9 cd         ( 2) 0.621570
10 rmail      ( 5) 0.567892
```

Example queries processed by PELICAN

```
I wish to change directory.

 1 cd          ( 2) 0.894543
 2 ls          ( 1) 0.828476
 3 mv          ( 4) 0.718574
 4 rmdir       (10) 0.717317
 5 more        ( 3) 0.702810
 6 pwd         ( 7) 0.692261
 7 rsh         ( 8) 0.689359
 8 rmail       ( 5) 0.654914
 9 mail        ( 6) 0.617851
10 rm          ( 9) 0.614234


I want to move some files

 1 mv          ( 4) 0.707107
 2 cd          ( 2) 0.457779
 3 pwd         ( 7) 0.398122
 4 ls          ( 1) 0.393605
 5 more        ( 3) 0.383323
 6 rm          ( 9) 0.366187
 7 rmdir       (10) 0.366187
 8 mail        ( 6) 0.342093
 9 rsh         ( 8) 0.337945
10 rmail       ( 5) 0.260635


I want to send some mail.

 1 mail        ( 6) 0.707107
 2 pwd         ( 7) 0.447774
 3 rm          ( 9) 0.423212
 4 rmdir       (10) 0.423212
 5 ls          ( 1) 0.414030
 6 rsh         ( 8) 0.411766
 7 cd          ( 2) 0.377377
 8 more        ( 3) 0.364211
 9 rmail       ( 5) 0.350016
10 mv          ( 4) 0.342093
```

More example queries processed by PELICAN

phrases and prepositional phrases. In our present scheme we are simply ignoring everything in the phrase except the head noun. Thus phrases such as 'a large file', 'many large files', and 'an interesting text file' are all considered to be synonymous with 'file'.

In concluding this section we should stress that the simple transformations which we are using at present are only a preliminary attempt at addressing a very difficult problem. We anticipate devoting a considerable amount of attention to this area in future.

4.3 Facts about the System and Preliminary Results

In this section we collate information about PELICAN and outline what results have so far been established. As we have already said, all distributed representations within the system are constructed in terms of three sets of features. The nine case features (shown in the table earlier) are closely related to the standard sets of semantic cases as developed by Fillmore and others. The other sets of features, the *object features* (for nouns) and the *action features* (for verbs) were developed for the PARROT project (Sutcliffe 1988a [18], 1991b [23]) and were used unchanged in PELICAN. There are 166 object features and 151 action features. The lexicon was re-coded for this project and at present contains around 100 words. The reason for recoding was simply that the words in PARROT's lexicon were suitable for discourse about domestic events — for example we had nouns like 'alarm', 'bicycle', 'coins' and 'dinner'. These are not very useful words in the context of the UNIX system.

The parsing process is at present accomplished by a grammar containing around thirty rules. This appears adequate for simple queries within a restricted and highly focussed domain, especially as it is designed to accept meaningful but nevertheless syntactically ill-formed strings such as 'want to move a file'.

At present we are working with a set of ten information files derived from the list used by McDonald and Schvaneveldt (1988) [27]. The propositions corresponding to these files have been hand coded using the representation as shown in the earlier 'ls' example.

Turning to the results so far, In the next two figures we show the output of the system for some example queries which PELICAN can process. As can be seen, the system outputs a list of all the information files it knows about and these are ordered by their perceived relevance to the query. The decimal number alongside each file name is the match of its corresponding proposition with the proposition into which the system transformed the query. This match varies between 0 (no match) and 1 (perfect match). As can be seen, the system performs well on the queries shown.

5 Summary and Conclusions

In this paper we have presented a preliminary experiment in the use of distributed propositional representations in a practical AI system called PELICAN. PELICAN can process simple queries about how to perform tasks on the UNIX system and can recommend appropriate information files which the user can consult in order to find out how those tasks may be accomplished.

PELICAN works by converting both queries and data about information files into a canonical semantic form. This form is a Fillmorian case structure where both semantic cases and their fillers are distributed patterns defined over sets of features. The advantage of using such a representation is that any sentence can be converted into a vector of fixed length which represents its meaning. Once we have meanings captured using such vectors the meanings can be compared very simply by comparing the vectors. For this task the standard vector comparison algorithms like the dot product can be used.

The main conclusion of this project is that the propositional representations appear to be quite effective in the chosen task. However the set of information files which the system knows about is at present very small (only ten commands). The next step, therefore, must be to expand this set to around one hundred commands, and to see whether the system still works. In addition we will then be able to measure the performance of the system using some more objective measures such as the *precision* and *recall* which are widely used for comparing IR systems (eg Salton 1986 [28]).

A major flaw with the present work (and indeed with almost all AI research) is that we are relying on *ad hoc* and subjective methods for performing tasks such as encoding lexical entries and constructing propositions corresponding to information files. Naturally this is undesirable although it is a useful approach for preliminary experiments. There are two ways in which lexical entries could be encoded in a more satisfactory manner. Firstly we could use some of the well-established techniques within psycholinguistics for eliciting lexical entry data from people (eg Katz 1983 [29]). Essentially this would amount to taking the hand coded patterns for a given lexical entry produced by different people and averaging them, thus reducing the subjective element. A second approach is to try to derive such representations from Machine Readable Dictionaries or other large databases of linguistic information. Workers such as Guo (1990) [30], Guthrie et al. (1989) [31] and Wilks et al. (1990) [32] have already made some useful progress in this direction.

Finally, turning to the matter of encoding propositions for information files, it is hoped that it will prove possible to generate these by scanning the text directly. There is of course a lot of work in this area already. For example in Smeaton's SIMPR project (Smeaton and Sheridan 1990 [13], Smeaton, Voutilainen and Sheridan 1990 [14]) sophisticated text scanning techniques are used to accomplish an exactly analogous goal, namely the encoding of text portions into Smeaton's Tree Structured Analytic (TSA) formalism. However, in our case we believe that the use of distributed representations will prove to have a number of advantages. For example, one of the operations which we can perform is to combine the meaning of a sentences simply by adding together their propositional representations. This is easy to do because the propositions are matrices. The efficacy of this technique was demonstrated in Sutcliffe (1988a) [15] where we showed that a 'gist' pattern for a knowledge structure could be generated by combining the patterns corresponding to constituent actions within that knowledge structure. This kind of property is likely to prove useful when the problem of text scanning is addressed because collating the representations corresponding to a large number of sentences into a single 'gist' representation is of course exactly what we wish to accomplish.

References

1. Bruce B. Case Systems for Natural Language. Artificial Intelligence 1975; 6:327-360

2. Katz JJ, Fodor JA. The structure of a semantic theory. Language 1963; 39 (2):170-210

3. Chomsky N. Aspects of the theory of syntax. MIT Press Cambridge MA,1965

4. Rosch E. Natural categories. Cognitive Psychology 1973; 4:328-350

5. Wilks Y. An intelligent analyser and understander of English. Communications of the Association for Computing Machinery 1975; 8 (18): 264-274

6. Lehnert WG, Burstein MH. The Role of Object Primitives in Natural Language Processing. Proceedings of the 6th IJCAI, 1979

7. Schank RC. Conceptual Dependency: A Theory of Natural Language Understanding. Cognitive Psychology 1972; 3 (4):552-630

8. Fillmore C. The case for Case. In: Bach E and Harms RT (eds) Universals of Linguistic Theory. Holt, Rinehart and Winston, New York, NY, 1968, pp 1-88

9. Sowa JF. Conceptual Structures, Information Processing in Mind and Machine. Addison-Wesley Reading MA London UK, 1984

10. Salton G. (ed). The SMART Retrieval System - Experiments in Automatic Document Processing. Prentice Hall Englewood Cliffs NJ, 1971

11. Croft WB. Approaches to Intelligent Information Retrieval. Information Processing and Management 1987; 23 (4):249-254

12. Smeaton AF. Information Retrieval and Natural Language Processing. In: Jones KP (Ed) Informatics 10. ASLIB London UK, 1989, pp 1-14

13. Smeaton AF, Sheridan P. Using Morpho-Syntactic Language Analysis in Phrase Matching Technical Report SIMPR-DCU-1990-169e, School of Computer Applications, Dublin City University, Glasnevin, Dublin 9, Republic of Ireland, 1990

14. Smeaton AF, Voutilainen A, Sheridan P. The Application of Morpho-Syntactic Language Processing to Effective Text Retrieval Technical Report SIMPR-DCU-1990-165e, School of Computer Applications, Dublin City University, Glasnevin, Dublin 9, Republic of Ireland, 1990

15. Lewis DD, Croft WB, Bhandaru N. Language-Oriented Information Retrieval 1989; International Journal of Intelligent Systems 4:285-318

16. Hahn U, Reimer, U Topic Essentials. Technical Report TOPIC-19/86, Universitaet Konstanz, FRG, 1986

17. Thiel U, Hammwohner U. Informational Zooming: An Interaction Model for the Graphical Access to Text Knowledge Bases. Tenth Annual International ACM SIGIR Conference on Research and Development in Information Retrieval, ACM Press, pp 45-56, 1987

18. Sutcliffe RFE. A Parallel Distributed Processing Approach to the Representation of Knowledge for Natural Language Understanding. Unpublished doctoral thesis, University of Essex, UK, 1988a

19. Sutcliffe RFE. Natural language paraphrase and the PARROT system (Abstract). First Annual Meeting of the International Neural Network Society (INNS-88), Boston, MA, 1988b

20. Waltz DL, Pollack JB. Massively parallel parsing: A strongly interactive model of natural language interpretation. Cognitive Science 1985; 9:51-74

21. Rumelhart DE, McClelland JL. On learning the Past Tenses of English Verbs. In: McClelland JL, Rumelhart, DE (eds) Parallel Distributed Processing: Explorations in the microstructure of cognition Volume II: Psychological and Biological models. MIT Press Cambridge MA, 1986, pp 216-271

22. Sutcliffe RFE. Distributed Subsymbolic Representations for Natural Language: How many Features Do You Need? Proceedings of the 3rd Irish Conference on Artificial Intelligence and Cognitive Science, 20-21 September 1990, University of Ulster at Jordanstown, Northern Ireland. Springer Verlag Berlin FRG Heidelberg FRG New York NY, 1991a, pp 279-305

23. Sutcliffe RFE. Representing Meaning using Microfeatures In: Reilly R, Sharkey NE, (eds) Connectionist Approaches to Natural Language Processing. Lawrence Erlbaum Associates Hillsdale NJ, 1991b

24. Smolensky P. A method for connectionist variable binding. TR CU-CS-356-87, February, Department of Computer Science, University of Colorado, Boulder CO 1987a

25. Dolan CP, Smolensky P. Tensor Product Production System: A Modular Architecture and Representation. Connection Science 1989; 1 (1):53-68

26. Pereira FCN, Warren DHD. Definite Clause Grammars for Language Analysis - a Survey of the Formalism and a Comparison with Augmented Transition Networks. Artificial Intelligence 1980; 13:231-278

27. McDonald JE, Schvaneveldt RW. The Application of User Knowledge to Interface Design. In: Guindon R. (Ed) Cognitive Science and its Applications to Human-Computer Interaction. Lawrence Erlbaum Associates Hillsdale NJ, 1986, pp 289-338

28. Salton G. Another Look at Automatic Text-Retrieval Systems. Communications of the Association for Computing Machinery 1986; 20 (2):648-656

29. Katz AN. Dominance and Typicality Norms for Properties: Convergent and Discriminant Validity. Behavior Research Methods and Instrumentation 1983; 15 (1):29-38

30. Guo C. Deriving a Natural Set of Semantic Primitives. Proceedings of the 2nd Irish Conference on Artificial Intelligence and Cognitive Science, 14-15 September 1989, Dublin City University, Ireland. Springer Berlin FRG Heidelberg FRG New York NY, 1990, pp 295-312

31. Guthrie L, Slator BM, Wilks Y, Bruce R. Is there content in empty heads? Proceedings of the 15th International Conference on Computational Linguistics (COLING-90) Helsinki, Finland. Association of Computational Linguistics Press, 1989, pp 138-143

32. Wilks Y, Fass D, Guo C, McDonald JE, Plate T, Slator BM. Providing Machine Tractable Dictionary Tools. Machine Translation 1990; 5:99-154

Student Model Refinement: A Case Study in CSCW.

Conn Mulvihill

Software Centre,
University College,
Galway

Gabriel McDermott

CAPTEC,
St. James Terrace
Malahide
Dublin.

Abstract

In this paper, we address the important issue of refining a student's user model. Our discussion is set in the context of a student browsing course databases. We discuss how CSCW provides an enabling vehicle for dialogue between the student and an adviser: Specifically, we show how through the use of dialogue focussing techniques, justified updates may be made to the student model.

1. Introduction

The traditional emphasis of work in computing in the area of dialogue has been on Human Computer Interaction (HCI) [1, 5, 11]. Models have been proposed that represent the interplay between a user and a system. However the question of focus in dialogue has previously been finessed. A good focus is essential for effective and efficient communications between the user and the system, and the provision of this focus requires a knowledge of the user's relevant characteristics. Recent work has begun to address this issue. and has drawn heavily on user modelling techniques[2, 4].

In the novel area of Computer-Supported Cooperative Work (CSCW) [3, 6, 10, 12], a computing system has to support dialogue between two or more users that wish to discuss an application. This is in addition to supporting the interaction between the users and the application, which is the emphasis of HCI. Now, not only must the question of focus between the users and the system in question be taken into account, but also the question of focus between the parties in the discussion [13].

We have developed methods for addressing this issue of focus in dialogue, (See also [9]). In this report, we present some considerations that motivated the work, and discuss the approach we have adopted to the problem. We illustrate our techniques with reference to an educational environment.

Specifically, in this paper, we show how a adviser and student may make use of a browsing facility in order to assist the process of refining the student's model (for background information see [7, 8]). We will assume that the adviser has responsibility for ensuring the guidance of the student is satisfactory. The communication between the student and the adviser in dialogue form is set in the context of the student's browsing activity. It is important to note that our discussion is within an educational environment. In such an environment, it cannot be assumed that the student's characterisation will either be correct or fixed at the start of affairs. Information may be lacking, or quite simply the characterisation may be inaccurate with respect to the goals of the student. Our work is aimed at discussing how the the characterisation of the student may be refined in a principled fashion.

Having outlined the theme of the discussion, we will now turn to consider the student model, and discuss the necessity for model refinement. We will subsequently illustrate our discussion with an example that presents the main features of our approach.

2. The Student Model

Information on the student forms the basis for many interactions between an educational service and the student himself. This information is to be found in the student model, which is meant to describe relevant characteristics of the student. Initially, when new users comes to make use of system facilities, information about those users will be quite sparse. Therefore the information must be obtained if students are to be dealt with in a satisfactory fashion by the system.

Such information can be acquired in various ways and at different stages. For example, certain administrative and educational information may be obtained immediately on joining the educational service. This will include the normal general details available on personnel files. Other information however may be obtained during the course of using the available services. For example, during the course of a browsing session, drawing on an on-line database of available courses, it may be possible to refine the current description of a student. It is this type of information that we wish to concentrate on in this paper.

We will consider in our discussion the interactions that occur between a student and a adviser when the student is engaging in browsing. We will assume that a browsing facility, an on-line database of courses, and communications between the adviser and the student are all available. The approach taken results in the student being provided with a higher quality of service through dialogue interactions between the adviser and the student. This will be because the adviser has been instrumental in guiding the student to sanction a more appropriate characterisation. The dialogues that occur between these two parties may be focussed by the knowledge gleaned through a perusal of the available course material, as found in the browsing request.

We see here the interaction of several educational components. The browsing facility is providing the context for evaluating the current partial description of the student, as found in the student model. The adviser is engaged in a dialogue with the student, attempting to provide guidance in the light of the student's responses to the courses retrieved. Thus we have here both the interplay between a student and an application, (the browsing facility), and the interplay between two user classes (student and adviser). This sets our discussion within the ambit of CSCW.

3.The Model Refinement Process

We will now present a first sketch of the model refinement process (see figure 1). We assume that the student has engaged in a browsing request, has examined the retrieved courses, and requires as a result to discuss the aptness of his current characterisation with the adviser. Note that we do not require the student to be in a position to indicate precisely why he considers his characterisation inappropriate. This will be discovered during the interaction with the adviser. All we require of the student is to contact the adviser in order to initiate a discussion on the current state of affairs.

We envisage that the adviser possesses considerably more functionality than the student in interactions between them. This results from the greater knowledge and experience of the adviser in the course selection process. For example, the adviser has the functionality to examine connections that the student will be unaware of between certain courses. This enhanced knowledge and functionality enables the adviser to guide the student to a more appropriate characterisation.

Figure 1

MODEL REFINEMENT STEPS:

1. Student submits a request to use the browsing facility.
2. Courses are retrieved based on the student's interests as stated in the student model.
3. Student evaluates retrieved material, and contacts adviser.
4. Dialogue between student and adviser results in the submission of a modified browsing request.
5. Continued dialogue may result in further submissions, or the updating of the student model.

As can readily be seen from Figure 1, we view model refinement as essentially a context-sensitive mechanism for updating the student model. In our sketch, the browsing facility provides the context of operation, and the adviser and student engage in a dialogue that draws on this context to improve the student's characterisation. We will turn now to consider general conditions of invocation, and subsequently we will examine an example that illustrates the exchanges that occur.

4. Invocation of Model Refinement

We consider that the inadequacy of the current student characterisation may be revealed in at least two ways. First, the student may discover, on the basis of a browsing session, that his characterisation seems inadequate, and calls upon the adviser for assistance. We have discussed this in the preceding section. Again, the adviser may ask the student to provide information that the adviser deems necessary. As an example of this, consider that a choicepoint may have arisen in the student's curriculum, and it is necessary for the student to select one of several courses. The student's adviser will have the responsibility of guiding the student to an appropriate choice, which will involve expanding the current description in the student model, if it is inadequate to uniquely determine a choice.

In both these cases, there is considerable scope for dialogue between the student and the adviser. The focus of this dialogue must be on providing a good refinement to the student's model. It is the adviser, on the basis of his enhanced knowledge and experience that will drive this process of refinement. We will now turn to consider further how this refinement is brought about. We will build on the sketch that has been established in section three.

5. Refining the Learner Model

In this section, we present an overview of one refinement process, and subsequently present an example that illustrates the method adopted. We then consider further possibilities that may arise. We concentrate on demonstrating how focussing techniques may be applied in the refinement process. We start with a simple case, in which the student actually is in a position to aid the adviser, although the student cannot explicitly determine this.

5.1 Simple Case Overview

Let us assume, therefore, that the student has browsed the courses available in the on-line database. He discovers that only some of the courses that have been retrieved hold an appeal for him. He notes those that he feels are in fact close to what his interests. He contacts the adviser.

Now these events lead to a dialogue between the adviser and the student which focuses on the noted courses in the context of the student model. The adviser will inspect these noted courses, using them to structure his exchange with the student. On the basis of this exchange, he will determine what he believes to be a more appropriate characterisation for the student. He then initiates a browsing request with this new characterisation, and the results of the search will be displayed to the student. This will provoke further exchanges, and further negotiations may result in an acceptable characterisation. The adviser will sanction the updating of the student model with any such acceptable characterisation.

Note here that a very focussed exchange has occurred. The browsing environment of operation provides the first focus. We have already restricted our attention to the student's academic interests. The noted courses provides the second. Here, we have an indication of the actual interests of the student. The third focus is provided by the adviser interacting with the student in response to these courses. This provides the student with the option to choose between alternative characterisations. These foci provide an excellent illustration of the uses of CSCW methods, with dependencies established between all parties in a discussion, and use being made of an application browsing facility.

5.2 Example of the Method

The student has noted courses in a browsing session. Let us suppose that he has noted mathematics related courses, whereas his model indicates that his interest is in computing. This may have occurred for a variety of reasons. For example, the initial characterisation may not have been accurate at that time. Again, it is possible that the interests of the student have changed over the course of time, and he feels that these courses reflect his current interests more closely. Or it may be quite simply that the student's true interests lie in the direction of formal computing methods.

The adviser has the functionality to explore the options that arise at this point. We will make the following two assumptions in our discussion. First, the student model is in fact in need of refinement, and the student is not in error. This assumption is easily checked by the adviser. Second, we assume the provision of a list of course descriptors. This is for illustration purposes.

The exchanges that occur are outlined in figures 2. The exchange between the adviser and the student occurs in a dialogue window that is separate from the window that carries the interplay between the users and the browsing application. This modularity in dialogue adds to the clarity of the process.

In the first case, the adviser concentrates on exploring the mathematical theme. He suggests that general mathematical courses may be of more interest to the student than computing courses. The student is prepared to provisionally accept this finding, and a browsing request is submitted with this characterisation on behalf of the student. Note that the student will observe the results of this request on his screen. He will therefore be enabled to examine the descriptions of these courses as before, and will respond to the adviser on the basis of his examination.

Figure 2

Search dialogue:	Adviser-Student exchange:
<Earmarked Courses>	Adviser: Let's try with the term 'mathematics', I think it is closer to what you have in mind.
	Student: OK
<New Courses>	

In the second case, the adviser would concentrate on exploring the formal computing theme, considering that formal computing courses may be of more interest to the student than the retrieved general computing courses. A browsing request is submitted with this characterisation on behalf of the student. Again, the student will observe the results of this request on his screen.

We see here that the exchange is focussed on the noted courses as a vehicle for suggesting changes to the student's current characterisation. If these new courses do actually reflect the student's interests, then the process will result in a more acceptable characterisation. Otherwise, the student will indicate which of the retrieved courses are closer to his interests, and the process can repeat in an iterative fashion.

5.3 Further Possibilities

However it may be the case that no courses have been noted as interesting by the student. Or it may be the case that the noted courses are not a good measure of the student's interests. Now if no courses have been noted, or they do not prove satisfactory, it is necessary to find another focus. If no focus were to be found, the only approach that could occur would be for a general selection of courses to be displayed to the student by the adviser, reflecting the divisions that occur in the available course list. This would assist the process of characterisation in a top-down fashion, and is a quite general approach. It does have the associated disadvantage however of being a long process. The more divisions that arise, the longer this process will take.

What is useful then is to obtain some focus that can be used in the dialogue exchange. This may be achieved by considering the current student characterisation. The adviser has the functionality to inspect connections that exist between certain areas that are not immediately visible. This provides the basis for selecting a dimension and allowing browsing requests along it.

Let us consider this further. Suppose that the student is characterised as interested in 'physics', and has noted no courses in a browsing session, but has notified the adviser that the current characterisation is unsatisfactory. In a more complete sense, suppose that the characterisation should be 'experimental physics'. Now although the current characterisation is known to be unsatisfactory, it is not immediately clear how to remedy the situation. But if the adviser, using his additional functionality, looks at the known connections, it is clear to him that a physics dimension exists, along which the student can be guided (see table 1).

Table 1

APPLIED	NATURAL	THEORETICAL
Geophysics	Experimental Physics	Mathematical Physics
Climatology		

Now the adviser will assist the student in migrating along this dimension. Therefore a display of applied and theoretical courses will be made available by the adviser to the student through submitting browsing requests using these terms as indices. The student might indicate a preference, and the skew in his interests from the middle ground will be established.

Therefore a focus has been established on the interests of the student. This focus enables dialogue between the adviser and the student to be more directed than was possible if only weak methods apply. The assumption that the neighbourhood of the current descriptor is worth investigating reduces the search space considerably, and provides another constraining focus on the exchanges. Only if it is not successful do weak methods have to be fallen back on, and even here account can be taken of the possibilities tried in the focussed dialogue exchanges.

6. Conclusion

We have discussed in this report the usefulness of focussed dialogue techniques in exchanges between the adviser and the student in the context of student model refinement. We have seen how certain focussing agents may be determined, and discussed how they may be used to aid in the process of characterising the student. We have indicated that the interplay between the browsing service and the student and the adviser gives considerable scope for motivating and justifying the updating procedure that occurs for the student's educational descriptor model. We have argued in favour of this approach both from an efficiency standpoint, in that less search is required, and from the standpoint of effectiveness, in that a principled exchange can occur between the student and the adviser, leading to more certainty in the model updating process.

7. References

[1] Allen, J.F., Frisch, A.M. and Litman, D.J., "ARGOT: the Rochester Dialogue System". In the Second National Conference on Artificial Intelligence, 1982.

[2] Benyon, D., Innocent, P. and Murray, D., "System Adaptivity and the modeling of Stereotypes". In H.J. Bullinger and B. Schackel (eds.) Human Computer Interaction. INTERACT'87, IFIP Elsevier Science, 1987.

[3] Card, Stuart and A. Henderson. *"A Multiple, Virtual-Workspace Interface to Support User Task Switching"*. In CHI + GI 1987.

[4] Cooper, M., "Interfaces that adapt to the user". In J. Self (ed.) "Artificial Intelligence and Human Learning: Intelligent computer-aided instruction". Published by Chapman and Hall, London, 1988.

[5] Hartson, Rex H., and Hix, Deborah "Human-Computer Interface Development: Concepts and Systems for its Management", ACM Computing Surveys, Vol. 21, No. 1 March 1989

[6] Johnson, B., Weaver, G., Olson, M., Dunham, R., "Using a computer-based tool to support collaboration: A field experiment". In the Proceedings of CSCW'86 (Computer Supported Cooperative Work) held in Austin, Texas, 1988.

[7] *Li Yim, Y. T., Mulvihill, C. "Browsing and Selection Service* Vol I: Specification" Project ACES(D.1008) in DELTA programme. RS/DS/211.

[8] Li Yim, Y. T., Mulvihill, C. "Browsing and Selection Service Vol II: Design Document of prototype". Project ACES(D.1008) in DELTA programme. RS/DS/212.

[9] McDermott, Gabriel, Mulvihill, Conn, and Patel, Ahmed, "Focussed Interfaces for Network Management", to appear in Joint SAIEE/CSSA International Symposium and Exhibition on Network Management, Pretoria, South Africa, May 1991.

[10] Mantei, M., "Capturing the Capture Concepts: A case study in the design of computer-supported meeting environments. In the Proceedings of CSCW'88 (Computer Supported Cooperative Work) held in Portland, Oregon, 1988.

[11] Pfaff, G.E, "User Interface Management Systems", Proceedings of IFIP/EG Workshop on UIMS, Spinger-Verlag, Seeheim, Federal Republic of Germany, 1983.

[12] Sorgaard, P., "A cooperative work perspective on use and development of computer artifacts". In 10th Information Systems Research Seminar in Scandinavia (IRIS) Conference Vaskivesi, Finland 1987.

[13] Verdejo, M. Felisa, "Modelling Human-Human communication in a Distance Learning Environment". In ITS-Seminar Aarhus Denmark, December 1990.

Section 4:

Parallel and Distributed AI

Natural Language Methodologies

Parallel Techniques for Image Processing and Artificial Neural Network Simulation

Hidenori Inouchi, Niall McLoughlin
Hitachi Dublin Laboratory,
Trinity College,
Ireland.
Email: hinouchi@vax1.tcd.ie

Abstract

The recent emergence of systems composed of multiple processing elements and memory units, and their associated models of computation promise to alleviate many of the limitations of conventional Von Neumann architectures. The implication of this to the field of Artificial Intelligence is twofold, Parallel systems offer both a significant increase in computing power/speed available, and a more natural physical architecture for implementing parallel solutions to A.I. problems. However, these systems are often extremely complex both from a conceptual (design) and practical (implementation) point of view. In this paper we will analyse various parallel methods and the considerations in using these for problem solving in the areas of image processing and artificial neural network simulation.

1. Introduction

The field of Artificial Intelligence is by its very nature multi-disciplinary, bringing together the disciplines of Psychology, Philosophy, Mathematics, Computer Science, Engineering, and more recently researchers in the fields of Neurobiology, Psychophysics, and related neural sciences. The digital computer had a great influence in the shaping of Artificial Intelligence theories in the past. At present most computers are based on the Von Neumann architecture, that is, a single Central Processor Unit (CPU) connected to a disjoint memory which holds both the data and program code. The limits of such an architecture are widely recognized: the speed of signal propagation, heat dissipation, and how quickly the CPU can retrieve instructions and data from the separate memory, and write back results to the memory. The development of new parallel architectures composed of multiple CPUs and memory units show great promise in alleviating some of the above problems. Many areas of A.I. are benefiting from these architectures, especially the area of real time image processing and machine vision. Current supercomputers are powerful enough to perform large scale computations required by many complex vision tasks, but they are expensive, and their architectures are not always suited to A.I. research problems. Many supercomputers, known as vector processors, rely on pipelined floating-point processors with very short cycle times to achieve their speed. More general parallel architectures provide a different approach to increasing system performance. These machines do not rely on the current 'state-of-the-art' in floating-point processors but instead make use of multiple general-purpose CPUs. Much investigation is still being carried out to develop efficient methods for using these machines, but the rapid growth in their power and flexibility makes them suitable for many complex domains.

Our study has been carried out on a Meiko Computing Surface Transputer system composed of sixty four INMOS T800 transputer. Such a system, which is available at a fraction of the cost of a supercomputer, can provide up to a peak performance of 400 VAX MIPS (million instructions per second), or 120 MFLOPS (million floating point operations per second). Each transputer has 2Mbytes of local memory, providing a truly distributed memory architecture. Occam is the 'native' language of the transputer. Based on the CSP [1] model of computation it provides constructs for the explicit definition of parallelism. For more information on the occam language see [2].

Within this paper we discuss the use of parallel programming in the areas of image processing and neural network simulation. Of particular interest to our on-going research is flexible, reusable, transputer programming. In the next section we discuss general parallel programming techniques, their real applications to image processing and neural network simulation, parallel implementation of neural network models, and finally we finish with our conclusions derived from this work.

2. Parallel Programming Techniques

A number of simple paradigms exist for exploiting parallelism in a given task and mapping this parallelism onto a series of processors. These paradigms consists of 'Region based or Geometric decomposition parallelism', 'Task farming parallelism', and 'Algorithmic parallelism'.

2.1 Region Based Parallelism

This occurs when the region over which the algorithm is to be computed is divided geometrically so that each processor is allotted a portion of the region to be modelled. Neighbouring processors are allocated geometrically bordering chunks of the region, thereby reducing latencies and congestion. Due to this, each processor is able to perform the algorithm on a greatly reduced region, and if bordering information is necessary, (in edge detection for example), queries can be passed independently to the bordering processors. This technique is used extremely efficiently in low level vision algorithms, for example in convolving an image with a gaussian, thresholding an image, or in simple edge detection algorithms. This is due

both to the local nature of the computation, and that information on the bordering regions may be requested in parallel to the application of the processing of the image data.
Let us take the example of mapping an image onto a simple transputer network (See Figure 1). We can use either pipeline or a torus structure to divide the work amongst the slave processors equally. In this example a single master process splits the image up into a number of chunks which is equal to the number of slave processors available, and then routes each chunk to a respective worker process. On receiving an image chunk each worker process should apply the image processing algorithm to its own local subset of the image region. In parallel each worker should communicaties with neighbouring worker processes. Finally when the above is finished each worker process should apply the implemented algorithm to the border regions. This overlap of local communication with the actual number crunching provides a very effective method for implementing many region based algorithms.

2.2 Task Farming Parallelism

This type of parallelism is extremely useful when a subsection of code or number of processes have to be run a large number of times with different parameter settings. The idea comes from having a master process which co-ordinates all the work in the process farm. If there is work to be done the master 'farms' out this work to idle slave/worker processes, which proceed to execute a job, and then return their result, at which stage the master may request them to begin executing a new job. This method involves the master process dynamically sending work to processes which are in an idle state. To work efficiently, load balance over the slave process farm has to be kept equal. However, the master process need not know at all times which processors are idle and which processors are working, nor have a means of addressing a job to any particular slave. Usually, the routing kernel on Master and Slave play an important role to balance the load. This method can make use of the same Master/Slave configuration as used previously in 'Region Based Parallelism'.

2.3 Algorithmic Parallelism

Algorithmic parallelism is often used by vector processors. These machines assign the various steps of a floating point multiply to a pipe of processes, each of which performs a given task on some data and then passes its result on to the next. For this type of parallelism to be successful the work load should be spread out as evenly as possible, with buffering of data at various stages of the computation if necessary. This method involves the user splitting up a task into a number of sequential steps and linking these steps in a pipeline of processors. The input is fed into the top of the pipeline and results arrive at the bottom of the pipeline. The user should expect a startup delay as it takes time before the first result arrives at the bottom of the pipeline. Again this method can make use of the same Master/Slave configuration as 'Region Based Parallelism' and 'Task Farming Parallelism'.

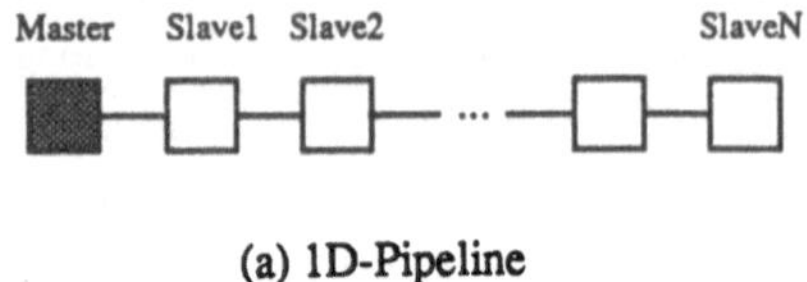

(a) 1D-Pipeline

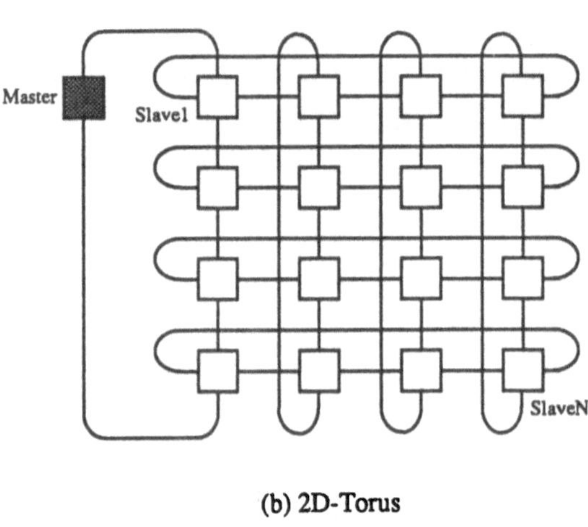

(b) 2D-Torus

Figure 1: Master/Slave Connection of Multiple Processors

3. HINTS - a case study

HINTS, or the Hitachi Neural Tracking System, was developed on 64 node transputer system [3]. HINTS is a real time target tracking system based on neural network technology. Target tracking has proven to be a very complex task for computer vision systems to achieve. A number of different approaches have been tried most of which require prohibitive amounts of computation for a standard single processor machine. Humans, on the other hand, find tracking targets to be, while not trivial, a relatively easily accomplished task. For this rational we designed HINTS using a model of the human visual system and basing the system on connectionist models. Although this approach normally requires a prohibitive amount of computation, vision is an inherently parallel activity, and so it is feasible to parallelise at least the low level image processing parts of a target tracking system. One of the major difficulties with tracking objects using camera data alone, is that it is possible for the object to become obscured by either, a second moving object (collision), or by a stationary object (occlusion).

We utilised the recall properties of various neural models to address these problems. Pre-attentative processes in human vision correspond to how areas of apparent motion or interest attract the human eye without the human being conscious of what is moving. In everyday life we experience this sort of mechanism when we catch something 'out of the corner of our eye'. The focus of attention process corresponds to us actually recognising the cause of interest by focusing our attention on the corresponding area.

Figure 2 shows a block diagram of HINTS. HINTS is split up into two functional blocks, the Interesting Point Locator Subsystem (IPLS), which corresponds roughly to the pre-attentative process, and the Motion Tracking Subsystem (MTS), which corresponds roughly to the focus of attention process of human vision [3]. The IPLS processes the images over time and using a simplified Motion Oriented Contrast (MOC) filter [3][4], coupled with a feature extraction module and a Centroid Detection Neural Network (CDNN), produces a series of coordinates which correspond to the rough centers of motion in the current image. This set of coordinates are used by the MTS as hints (hence the name) for zooming in on the regions of interest in the latest image (or focusing attention). This MTS determines whether a target is present or not.

The simplified MOC filter and the CDNN are the most computationally intensive parts of the system, and accordingly, up to 56 Transputers were used in their implementation.

3.1 Motion Oriented Contrast (MOC) filter

The MOC filter is a neurally based model which imitates certain cells in the retina and visual cortex to produce a measure of apparent motion between images which is similar in nature to the optical flow based calculation. The MOC filter outputs an approximate measure of strength of motion for each pixel in each of eight possible directions. The algorithm is composed of a number of steps each of which requires only local computation.

The best method to implement the MOC filter on multiple processors was found to be 'Region Based Parallelism', since purely local operations over a small region are needed in the calculation.

3.2 Centroid Detection Neural Network (CDNN)

The CDNN, which was derived from a modified instar/outstar competitive network [3][5], is used to automatically contrast enhance and noise suppress these input feature vectors until the centroids of motion become apparent. The CDNN is capable of recalling recent motions and is used in this manner to overcome problems of occlusion and collision. Finally these hints or cues are used by the system to determine when MTS 'looks' for the moving targets.

The best method to implement the CDNN on multiple processors was found to be 'Task Farming Parallelism', since a fairly large amount of global data communication is needed in the calculation.

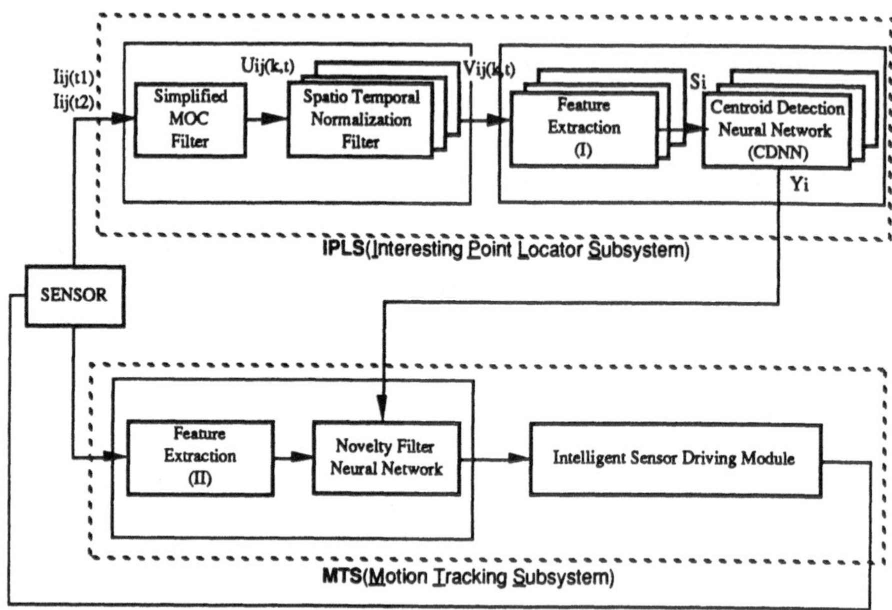

Figure 2: Functional Block Diagram of HINTS

4. Simulation of Neural Network Models

The following sections discuss the parallel implementation of neural networks. As an example, dynamic type of neural network, ART1 is used. Note that the network topology of CDNN is a subset of Augumented ART1(A-ART1) Neural Network.

It is often convenient to describe a variety of neural network models as dynamic systems. These dynamic systems may be either open loop or closed loop systems with feedback.

The occam language is ideally suited to describe dynamic systems which consist of concurrently working element blocks and therefore provides a natural mechanism for implementing neural networks. Each element block which defines the behaviour of the network can be described as a occam thread and a collection of such threads can describe total network behaviour in a natural way.

4.1 The Augmented ART1 (A-ART1) Model

The augumented ART1 model is a modified version of ART1 neural network model that facilitates its real-time implementation [7]. Figure 3 shows the block diagram of this model which has 3 layered structure. F0 depicts the external input layer, F1 depicts the comparison layer and F2 depicts the coding layer. The connection from F1 to F2 is termed 'Bottom-Up' and the connection from F2 to F1 'Top-Down'. The connection topology of the A-ART1 model is a superset of the typical 3 layered Backpropagation model and 1 layered Hopfield model. This being the case, the same parallel technique used to implement the A-ART1 model can be applied when implementing the Backpropagation or Hopfield model on multiple processors.

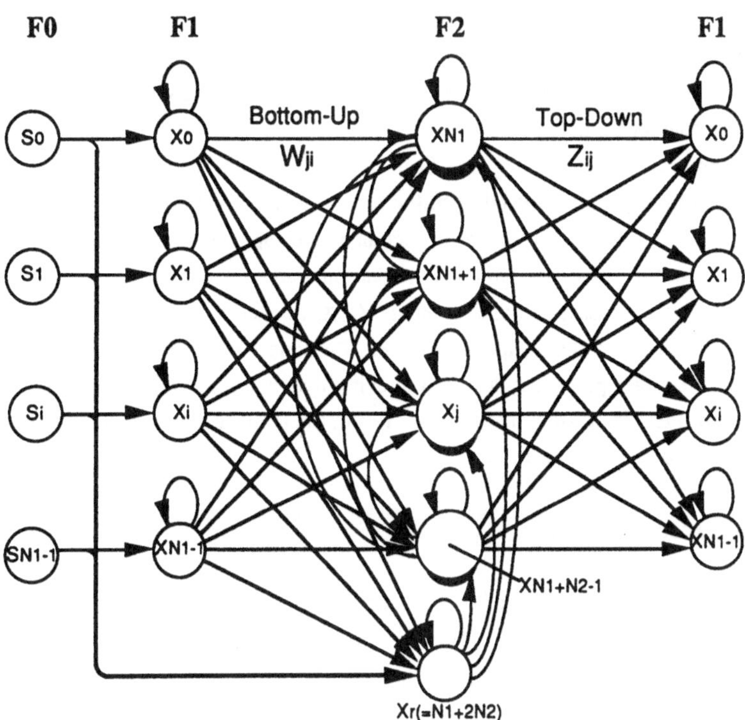

Figure 3: A-ART1 Neural Network Model

4.2 Implementation on a Single Transputer

Simulation code for A-ART1 in occam on a single transputer is shown in Figure 4. In this figure and figure 3, N1 and N2 represents the number of nodes belonging to the F1 and F2 layer, respectively. S_i represents ith element of external input vector. X_i represents ith node activation value in the F1 layer and X_j jth node activation value in the F2 layer. X_r represents reset node activation value. W_{ji} represents 'Bottom-Up' weight and Z_{ij} represents 'Top-Down' weight. Note that black circles behind the white circles in the F2 layer represent inhibition nodes [7].

Below, each process is described in detail.

4.2.1. Update Process

In general, each **Update** process is responsible for receiving inputs from the **Fowarder** process, updating its activation value based on its own update rule, and sending back the updated activation value to the **Forwarder** process. For example, if N1=N2=10, then 31 UpdateNode processes and 200 UpdateSynapse processes are instantiated by the **PAR** construct within the **Update** process. These processes run concurrently, using the on-chip hardware scheduler of the transputer.

Specifically, external inputs (training data S_i to each node in the F1 and F2 layer) are supplied via the channel from the **StimulusDistributor** process and internal inputs (lateral inhibition term and weighted sum of inputs term) to each node are supplied via the channel from the **ResponseCollector** process. Updated activation values are also sent back to the **ResponseCollector** process. Similarly, inputs X_i, X_j to each synapse in the 'Bottom-Up' and 'Top-Down' connection are supplied from the **ResponseCollector** process.

Note that each **Update** process is designed to exit only when it receives a terminate signal from the **StimulusDistributor** process. As a result, the number of repetitions of update is implicitly specified in that terminate signal.

4.2.2. Forwarder Process

The **Forwarder** process consists of two concurrent processes : the **StimulusDistributor** process and the **ResponseCollector** process. As mentioned above, the **StimulusDistributor** supplies external input to the network and the **ResponseCollector** process collects and distributes new activation values. For this purpose, **StimulusDistributor** process has a pool of storage for training data and **ResponseCollector** process has a large pool of storage for node and synapse activation values. In this way, synchronization between each instantiated **Update** process is automatically maintained, due to the channel communication between each **Update** process and the **Forwarder** process. Note that **ResponseCollector** process always keeps the latest activation values and thus it is easy to visualize network activation.

```
PROC Update()
  PAR
    PAR i=0 FOR N1
      PAR
        UpdateNodeF1(i)                    -- Update Xi
        PAR j=N1 FOR N2
          UpdateSynapseTopDown(i, j)       -- Update Zij
    PAR j=N1 FOR N2
      PAR
        UpdateNodeF2(j)                    -- Update Xj
        PAR i=0 FOR N1
          UpdateSynapseBottomUp(j, i)      -- Update Wji
    PAR j=(N1+N2) FOR N2
      UpdateNode_F2(j)                     -- Update _Xj
    UpdateNodeR(N1+2N2)                    -- Update Xr
:

PROC main()
  PAR
    Update()              -- Update Network Activities
    {{{ Forwarder()
    StimulusDistributor() -- Distribute stimuli to Update() procedure
    ResponseCollector()   -- Collect Network Activities and Distribute them to Update() procedure
    }}}
:
```

Figure 4: Multi Threaded Occam Code of A-ART1 on a single transputer

4.3 Implementation On Multiple Transputers

Now let us consider the way in which we can port the above occam program to a multiple transputer system to accelerate the execution speed. Unfortunately, the **PAR** construct in occam works only within a single program. Hence, it is not straightforward to port this program to a multiple transputer system as it stands.

The central aim of the simulation is to provide the flexibility necessary to allow different neural models to be incorporated cleanly. To date, almost all distributed neural implementations have been designed specifically for one neural model [8][9], and are by their very definition restrictive. The solution which we are pursuing is to try to remove the physical hardware considerations from the neural network designers code.

We will examine two methods of porting this program to a multiple transputer systems. One is based on 'Task Farming Parallelism' and the other is based on 'Region Based Parallelism'. To date the latter method has been more extensively studied [10] than the former. Both methods use a load balancing technique to achieve the most efficient use of multiple processors. The primary difference between the two methods is that 'Region Based Parallelism' allocates tasks

staticaly, whereas 'Task Farming Parallelism' allocates tasks dynamically. Figure 5 shows the occam code based on 'Task Farming Parallelism'.
Below, each process is described in detail.

4.3.1. SendTask Process

The **SendTask** process on the master processor sends out a series of packets to the **Forwarder** process. These packets request each slave processor to update the activation value (which is included in the packets) according to the type of the **Update** procedure.
Note that each **SendTask** process is designed to exit only when it receives a terminate signal from the **Forwarder** process.

4.3.2. Fowarder Process

The **Forwarder** process is a routing kernel on the master processor. It sends out packets sent from the **SendTask** process to each slave processor and concurrently collects the packets from each slave processor and then distributes them to the **SendTask** process.

4.3.3. Taskscheduler Process

The **Taskscheduler** process is a routing kernel on the slave processor. It receives packets from the master processor and then sends them to the **ReceiveTask** process. Meanwhile, it concurrently sends back the packets (which include updated activation values) sent from the **ReceiveTask** process to the master processor. Also, the **Taskscheduler** process serves to dynamically balance the load between each slave processor.

4.3.4. ReceiveTask Process

The **ReceiveTask** process is responsible for updating the activation value. It receives packets from the **Taskscheduler** process and based on the type and index number of the update (which is extracted from the packets) executes the relevant update.
Note that **ReceiveTask** process corresponds to the **Update** procedure in figure 4.

```
PROC SendTask()
  PAR
    PAR i=0 FOR N1
      PAR
        SendTaskUpdateNodeF1(i)                    -- Update Xi
        PAR j=N1 FOR N2
          SendTaskUpdateSynapseTopDown(i, j)       -- Update Zij
    PAR j=N1 FOR N2
      PAR
        SendTaskUpdateNodeF2(j)                    -- Update Xj
        PAR i=0 FOR N1
          SendTaskUpdateSynapseBottomUp(j, i)      -- Update Wji
    PAR j=(N1+N2) FOR N2
      SendTaskUpdateNode_F2(j)                     -- Update _Xj
    SendTaskUpdateNodeR(N1+2N2)                    -- Update Xr
:

PROC  main()          -- Master main procedure
  PAR
    SendTask()     -- Sends Tasks to Forwarder and Receives Results from Forwarder
    Forwarder()    -- Sends Tasks to Slaves and Receives Results from Slaves
:
```

(a) Occam Code for A-ART1 Master Processor

```
PROC ReceiveTask()
  SEQ
    . . . Receive Packets (type and index i,j  are extracted from the packets at this point)
    IF
      (type = UpdateNodeF1)
        UpdateNodeF1(i)                  -- Update Xi
      (type = UpdateSynapseTopDown)
        UpdateSynapseTopDown(i, j)       -- Update Zij
      (type = UpdateNodeF2)
        UpdateNodeF2(j)                  -- Update Xj
      (type = UpdateSynapseBottomUp)
        UpdateSynapseBottomUp(j, i)      -- Update Wji
      (type = UpdateNodeF2)
        UpdateNode_F2(j)                 -- Update _Xj
      (type = UpdateNodeR)
        UpdateNodeR(N1+2N2)              -- Update Xr
      TRUE
        SKIP
:

PROC main()                 -- Slave main procedure
  PRI PAR
    Taskscheduler()         -- Receives Task from Master and Sends Results to Master
    ReceiveTask()           --  Receives Task from Taskscheduler and Sends Results to Taskscheduler
:
```

(b) Occam Code for A-ART1 Slave Processor

Figure 5: 'Task Farming Parallelism' of the A-ART1 on Multiple transputers

The primary advantage of using 'Task Farming Parallelism' is that a user is able to write different update modules, corresponding to different neural network models, and incorporate these cleanly into the existing system. An additional advantage is that the modules are relatively hardware independent, (especially if a routing kernel is incorporated), allowing extra slave processors to be added to the system without the need to rewrite the distributor and routing code.

5. Conclusions

Recently great improvements have been made in the development of parallel hardware systems. However much work still needs to be done if these systems are to be utilised effectively. We have described a number of parallel programming techniques, and how these techniques were applied to a complex machine vision task. Much of the development time in constructing HINTS was spent in designing and implementing routing and distribution software. We believe that the idea of separating the routing and distribution software from the actual application software, to allow greater flexibility in hardware resources and easier code usability, will prove important if parallel systems are going to become more accepted by the general AI community. Many researchers would not be prepared to spend a significant proportion of time working on the routine aspects of parallel implementation such as routing and distribution software. An example of one method to overcome this is the routing kernel Tiny2 used by the Edinburgh Concurrent Supercomputer Program [11]. This allows the same routing kernel to be used no matter what form of parallelism is required. We feel that systems such as this will prove most useful in general AI research in the near future.

Acknowledgements

Many thanks to our manager Nobuo Hataoka for his support throughout this research project and Mark Whitehouse for his useful suggestions.

References

[1] Hoare, C.A.R., "Communicating Sequential Sequential Processes", *Communications of the ACM*, Vol 21, No. 8, 1978, pp 666-677.

[2] Hoare, C.A.R., Ed., "Occam2 Reference Manual", INMOS Limited, Prentice Hall, 1988.

[3] Grossberg, S., Rudd, M., "A neural architecture for visual motion perception : Group

and element apparent motion", *Neural Networks*, Vol 2, 1989, pp 421-450.

[4] Inouchi, H., McLoughlin, N., & Vernon, D., "A real-time simulated human vision system using connectionist models applied to target tracking", *The Proceedings of the IAPR Workshop on Machine Vision Applications*, 1990, pp 303-306.

[5] Carpenter, G., "Neural network models for pattern recognition and associative memory", *Neural Networks*, Vol 2, 1989, pp 243-257.

[6] Grossberg, S., "Nonlinear neural networks : Principles, mechanisms, and architectures", *Neural Networks*, Vol 1, 1988, pp 17-61.

[7] G.L.Heileman and M.Georgiopoulos. "The Augumented ART1 Neural Network", *Proceedings of the IJCNN*, July 1991, volII, pp467-472.

[8] Iazzetta, A., Vaccaro, R., & Villano, U., "A transputer implementation of boltzman machines", *Parallel architectures and Neural Networks*, 1988, pp 128-145.

[9] Richards, G., Tollenaere, T., "A revised version of the Rhwydwaith neural net simulator", *ECSP-UG-7*, 1989.

[10] Y.Fujimoto, N.Fukuda, "An Enhanced Toroidal Lattice Architecture Neurocomputer for Large Scale Neural Networks", *Proceedings of the IJCNN*, June 1989, volII, pp 614.

[11] Clarke, L., Wilson, G., "Tiny: An effecient communications harness for the INMOS transputer", *ECSP-UG-9*, 1989.

Topicalisation and Attachment Preferences

Allan Ramsay

University College, Dublin

Abstract

At first sight, "topicalisation" — the process of moving some constituent of a sentence to the front for emphasis — seems to be a fairly uncommon syntactic phenomenon. We argue that it is far more prevalent than is usually recognised, and that in a number of situations it leads to a reduction in syntactic ambiguity and hence to easier processing. This reduction in ambiguity can be used to support a principled defence of certain "preferences" in the attachment of modifiers such as PP's and adjectival VP's.

1 Topicalisation of Core NP's

Consider the following sentence (which should be read with emphasis on the words *bike* and *stole*, otherwise it won't sound like a sentence at all):

(1) *My bike he stole.*

It is generally argued that this is closely related to

(2) *He stole my bike.*

but that (1) would be used in situations where the speaker wanted to emphasise what was stolen. It is argued that (1) and (2) are related by a process of "topicalisation" whose underlying syntactic description is very similar to that of the process "WH-movement" which leads to relative clauses such as:

(3) ... *which he stole* ...

We follow common practice in unification grammar in describing both these phenomena in terms of a category-valued feature called `slash` which can be used for copying the syntactic and semantic properties of a displaced item to the appropriate place. Schematically the rule for both topicalisation and WH-movement looks like:

$$S(\ldots) \longrightarrow X(\ldots), S(\ldots, slash(X), \ldots)$$

In other words, you can make a sentence out of an item of type X followed by a sentence which has an item of type X "missing", in which case the details of the initial item are copied to the place where something of that kind is required. For further details see any standard text on unification grammar (e.g. [Pereira & Warren 1980, Gazdar et al. 1985, Shieber 1986]). Within the framework described in [Ramsay 1990a] this leads to the following semantic analysis of both (1) and (2) (all the semantic analyses in this paper are generated by the program described in [Ramsay 1990a]):

```
PRESUPPOSITIONS:

male(A)+speaker(B)+own(B,C)&bike(C)

PROPOSITION:

exists(D)action(D,steal)&(agent(D,A)&object(D,C))
        & exists(E):{interval(E)}
                exists(F):{instant(F)}
                     before(F,now)&contains(E,F)
                & during(E,D)
```

The presuppositions here say that the relevant sentence will only be interpretable in discourse situations where there is a uniquely identifiable male individual (corresponding to the pronoun *he*), a uniquely identifiable speaker (corresponding to the possessive determiner *my*) and a uniquely identifiable bicycle belonging to the speaker (from the NP *my bike*). The proposition then increments the discourse situation by introducing a past stealing event whose agent is the male individual and whose object is the bicycle. For further details of the relations between presuppositions and propositions see [Ramsay 1990a, 1990b]. For the purposes of the current paper we will simply assume that this kind of formal paraphrase can be given a rigorous semantics which corresponds reasonably well to the normal interpretations of (1) and (2).

Out of context, (1) looks rather stilted. This is hardly surprising. (1) is an alternative to (2), to be used in situations where the object which was stolen is to be emphasised. Without any surrounding context it is very hard to see why a particular item should be emphasised, and hence it is hard to see why (1) would be chosen rather than (2). The following example should be more convincing:

(4) *I liked the pear but the peach I thought was inedible.*

PRESUPPOSITIONS:

```
speaker(A)+pear(B)+speaker(C)+peach(D)
```

PROPOSITION:

```
exists(F)state(F,like)&(agent(F,A)&object(F,B))
         & exists(G):{interval(G)}
                 exists(H):{instant(H)}
                       before(H,now)&contains(G,H)
                 & during(G,F)
& exists(I)state(I,think)
           & agent(I,C)
           & object(I, *[exists(J)condition(J,inedible,K)
                                   & object(J,D)
                                   & exists(L):{interval(L)}
                                           exists(M):{instant(M)}
                                                 before(M,now)
                                                 & contains(L,M)
                                           & during(L,J)])
           & exists(N):{interval(N)}
                   exists(O):{instant(O)}
                         before(O,now)&contains(N,O)
                   & during(N,I)
```

The presuppositions correspond to the two definite NP's *the pear* and *the peach* and the two instances of *I*. The two variables *A* and *C* should, of course, get bound to the same individual by any subsequent inference algorithms, but there is nothing in the initial semantic analysis to enforce this.

The analysis above introduces two pieces of notation which, while not crucial to the argument in this paper, probably deserve explication. The first is the use of restricted quantifiers in expressions of the form `exists(X):{R}P`. Such an expression should be read as "for some X such that R holds of X, P also holds of it". This is an entirely standard convention. Formally `exists(X):{R}P` is equivalent to `exists(X)(P & Q)`. The notation `*[...]` is less orthodox. Its function is to make it possible to treat propositions as individuals, capable of entering into the same kinds of relations as other individuals. Thus the object of the speaker's state of thought is the proposition that the peach was inedible. Turner [1990] provides a comprehensive review of the properties of such intensional objects.

The interesting thing about (4) for the current paper is the topicalisation of the NP *the peach.* (4) is related to

(5) *I liked the pear but I thought the peach was inedible.*

in the same way that (1) is related to (2). In (4) it is the subject of the S-complement of *thought* which has been topicalised — fair enough, there are no constraints to rule out topicalisation of embedded subjects. (4), however, provides its own context for the emphasis of this item. The speaker has said something positive about the pear, and he or she now wants to contrast this with a negative comment about the peach (the use of *but* rather than *and* also contributes to this contrast. Apart from these instructions to the hearer about the speaker's attitude to what he or she is saying, there is no difference between *and* and *but*, which we therefore treat identically as far as the presuppositions and proposition of a text are concerned).

(1)/(2) and (4)/(5) illustrate the way that topicalisation of NP's from arbitrary syntactic contexts can be used to emphasise one item at the expense of others. In the remainder of this paper we consider topicalisation of other kinds of syntactic item, showing that the single rule for topicalisation given above can be used to provide simple analyses of constructions which would otherwise require a wide range of extra grammar rules.

2 Topicalisation of Other Items

The topicalised NP's in (1) and (4) are "core" arguments of their main verbs, in the sense described in [Foley & van Valin 1984]. Roughly speaking, core arguments provide fillers for thematic roles without which the VP in which they appear will be semantically incomplete. The same notion is seen in the allocation of obligatory roles to case frames in case grammar [Fillmore 1968], and in numerous other accounts. Topicalisation is not, however, restricted to core NP's. Consider for instance the VP *sitting in the park* in:

(6) *I saw an old man sitting in the park.*

One reading of this has the following formal interpretation:

```
PRESUPPOSITIONS:

speaker(A)+park(C)

PROPOSITION:

exists(D):{man(D)
          & exists(E)condition(E,old,*[F,man(F)])
                     & object(E,D)}
```

```
exists(G)action(G,see)
        & agent(G,A)
          & object(G,
                  *[exists(H)state(H,sit)
                              & agent(H,D)
                              & in(H, C)])
        & exists(I):{interval(I)}
                exists(J):{instant(J)}
                     before(J,now) & contains(I,J)
                & during(I,G)
```

The presuppositions say that (6) will only be meaningful in situations where there is a unique identifiable speaker and a unique identifiable park. The proposition introduces a past seeing action whose object is the proposition that the old man (the entity D which is a man and is the object of the condition E of being "old for a man") was sitting in the park. The schematic rule for topicalisation above suggests that

(7) *Sitting in the park I saw an old man.*

should be acceptable and should also have this interpretation, since the relation between (6) and (7) is exactly the same as that between (1) and (2). We note in passing that this interpretation is obtained using a sense of the verb *see* which has a VP as a core argument.

(6), however, also has a reading under which it was the speaker who was sitting in the park:

PRESUPPOSITIONS:

```
speaker(A)+park(C)
```

PROPOSITION:

```
exists(D):{man(D)
          & exists(E)condition(E,old,*[F,man(F)])
                    & object(E,D)}
     exists(G):{(state(G,sit) & agent(G,A)) & in(G, C)}
          exists(H)action(H,see)
                  & agent(H,A) & object(H,D)
                  & exists(I):{interval(I)}
                          exists(J):{instant(J)}
                               before(J,now) & contains(I,J)

                          & during(I,H)
                  & simultaneous(H,G)
```

The proposition of this interpretation refers to a sitting in the park, whose agent is the speaker, and an action of seeing whose agent is the speaker and whose object is the old man. This reading is obtained by interpreting the VP *sitting in the park* as a VP modifier on the main VP *saw an old man.*

This analysis of (6), then, involves the standard $S \longrightarrow NP, VP$ rule, where the VP in question is made up of a simple main VP and a VP modifier. Our rule about topicalisation says nothing about the role that the topicalised item should play, so there is no reason why VP modifiers of this kind should not get topicalised. We thus find that this interpretation is also available for (7).

(6) has yet more interpretations. Two of them arise from basically the same source as the ones we already have, except that instead of treating *sitting in the park* as a single VP we treat it as a VP *sitting* and a PP *in the park*. The first of these interpretations amounts to saying that I saw an old man sitting, and that where I saw him doing this was in the park. The second says that while I was sitting, I saw in the park an old man:

```
PRESUPPOSITIONS:

speaker(A)+park(C)

PROPOSITION:

exists(D):{man(D)
           & exists(E)condition(E,old,*[F,man(F)])
                      & object(E,D)}
      exists(G)action(G,see)
               & agent(G,A)
               & object(G, *[exists(H)state(H,sit) & agent(H,D)])
               & exists(I):{interval(I)}
                       exists(J):{instant(J)}
                             before(J,now) & contains(I,J)
                       & during(I,G)
               & in(G, C)
```

```
PRESUPPOSITIONS:

speaker(A)+park(C)

PROPOSITION:

exists(D):{man(D)
           & exists(E)condition(E,old,*[F,man(F)])
                      & object(E,D)}
      exists(G):{state(G,sit) & agent(G,A)}
            exists(H)action(H,see)
                     & agent(H,A) & object(H,D)
                     & exists(I):{interval(I)}
                           exists(J):{instant(J)}
                                 before(J,now) & contains(I,J)
                           & during(I,H)
                     & simultaneous(H,G)
                     & in(H, C)
```

The final interpretation is obtained by treating *sitting in the park* as a WHIZ-deleted relative clause. Under this interpretation what I saw was an old man who was sitting in the park:

```
PRESUPPOSITIONS:

speaker(A)+park(C)
```

PROPOSITION:

```
exists(D):{man(D)
          & exists(E)condition(E,old,*[F,man(F)]) & object(E,D)
          & exists(G)state(G,sit) & agent(G,D) & in(G, C)}
     exists(H)action(H,see)
              & agent(H,A) & object(H,D)
              & exists(I):{interval(I)}
                     exists(J):{instant(J)}
                          before(J,now) & contains(I,J)
                     & during(I,H)
```

It is fortunate that simple intransitive VP's cannot function in this way, so that *a man sitting* is not, at least to my intuitions, an acceptable NP in the same way that *a man sitting in the park* is (Consider, for instance, the relative acceptability of *An old man sitting in the park saw me* and *An old man sitting saw me*). If it were, we would find yet more possible interpretations for (6). As it is we have rather more than we would like.

These last three interpretations, however, are not available for (7). The first two are blocked because the rule for topicalisation says that *one* item may be extracted. If we wanted to deal with *sitting in the park* as two separate entities, the VP *sitting* and the PP *in the park* we would have to be able to topicalise *two* objects out of the same clause, which is simply not allowed. The third is blocked simply by insisting that nothing can be topicalised from the relative clause, which is a very standard constraint.

Thus the rule for topicalisation given above does two things for us. It provides a simple description of the way that (7) gets its interpretations, by treating it as though it were (6) but with the VP *sleeping in the park* moved/copied to the front. And rather more surprisingly it leads to less ambiguity for (7) than for (6), partly via the constraint that only one item may be topicalised (so that *sleeping in the park* has to be treated as a single VP rather than a VP and a PP) and partly via the constraint that objects cannot be topicalised out of relative clauses.

Similar considerations apply when we look at

(8) *I saw the old man in the park with a telescope.*

and

(9) *In the park with a telescope I saw the old man.*

(8) is notorious for its ambiguity. Was the old man in the park with a telescope, or is that where I saw him? Did I see him with the telescope, or is that just something that was in the park? Or was it perhaps something he had with him? And so on and so on. The two constraints on topicalisation, that only one item may be topicalised and that items may not be topicalised out of NP modifiers, mean that only the following interpretation survives as an analysis of (9):

PRESUPPOSITIONS:

```
speaker(A)
+ man(B) & exists(C)condition(C,old,*[D,man(D)]) & object(C,B)
+ park(F) & exists(G):{telescope(G)}with(F,G)
```

PROPOSITION:

```
exists(H)action(H,see) & (agent(H,A) & object(H,B))
        & exists(I):{interval(I)}
                exists(J):{instant(J)}
                     before(J,now)
                     & contains(I,J)
                     & during(I,H)
        & in(H,F)
```

This is the interpretation which presupposes the existence of a speaker, and old man and a park with a telescope, and provides the new information there was a seeing event which took place in the park, whose agent was the speaker and whose object was the old man.

3 Preferences

We have demonstrated so far that by adding a single rule for topicalisation to our grammar we get interpretations of numerous sentences which would otherwise each have required its own explication. More than that, we have shown that in several cases the topicalised version of a sentence is intrinsically less ambiguous than the non-topicalised version. We now suggest a tentative explanation of the reason why some of the interpretations of the non-topicalised version are more awkward sounding than others:

> Suppose I and I' are alternative interpretations of some sentence S. Suppose that I is also an interpretation of some sentence $top(S)$ which is obtained by from S by topicalisation, but that I' is not. Then
>
> I' should be preferred as the interpretation of S.

The justification for this rule is simple. Suppose I want to convey I' to you. My only choice is to utter S. I know that it is potentially ambiguous, but it's all I've got so I'll have to use it. Suppose on the other hand I want to convey I. I can convey this via either S or $top(S)$. It is clear that $top(S)$ is a better choice, since it is unambiguous so there is no chance of your getting confused about what I mean.

You as the hearer can now reason as follows. Suppose you hear me say S. You realise that I might want to convey I or I might want to convey I'. But you also know that if I had wanted to convey I' I would probably have said $top(S)$, which is unambiguous. So since you know I have a better way of conveying I', but no better way of conveying I, you conclude that I is likely to be the interpretation I wanted you to get.

This argument is not watertight. Various semantic preferences, of the kind often modelled using selection restrictions, can override it, as can purely stylistic considerations. Nonetheless, as a justification for heuristics such as left-attachment it does seem more convincing than theories which appeal to properties of particular processing strategies. In particular it provides some explanation for the fact that topicalised core NP's (as in (1)) seem far less fluent than other kinds of topicalised items. Our rule for attachment suggests that one reason for using topicalisation is to pick out a particular interpretation from among a number of competitors. A sentence like (2), however, is perfectly unambiguous anyway. We would therefore not use (1) in order to disambiguate it, so the only reasons for choosing (1) rather than (2) are associated with discourse matters such as topic and emphasis rather than simple low-level syntactic disambiguation.

REFERENCES

Fillmore C. (1968): The Case for Case, in *Universals in Linguistic Theory* (eds. E. Bach & R.T. Harms): Holt, Rinehart & Winston, Chicago: 1–90.

Foley W.A. & Van Valin R.D. (1984): *Functional Syntax and Universal Grammar*: Cambridge University Press, Cambridge.

Gazdar G., Klein E., Pullum G. & Sag I. (1985): *Generalised Phrase Structure Grammar*: Basil Blackwell, Oxford.

Pereira F.C.N. & Warren D.H.D. (1980): Definite Clause Grammars for Language Analysis — a Survey of the Formalism and a Comparison with ATNs, *Artificial Intelligence* 13(3): 231–278.

Ramsay A.M. (1990a): *The Logical Structure of English: Computing Semantic Content*: Pitman, London.

Ramsay A.M. (1990b): Presuppositions and WH-clauses, *DANDI workshop on presupposition*, Nijmegen.

Shieber, S.M. (1986): *An Introduction to Unification-based Approaches to Grammar*: Chicago University Press, Chicago.

Turner R. (1990): *Truth and Modality for Knowledge Representation*: Pitman, London.

An emergent computation approach to natural language processing

Jon Rowe
Department of Computer Science
University of Exeter
GB- Exeter EX4 4PT, E.C.
E-mail: jro@dcs.exeter.ac.uk

Paul Mc Kevitt
Visiting Fellow
School of Electronic Engineering
Dublin City University
IRL- Dublin 9, Dublin, E.C.
E-mail: mckevittp@dcu.ie

Abstract

One of the most difficult problems within the field of Artificial Intelligence (AI) is that of processing language by computer, or *natural-language processing*. Approaches to natural-language processing have been formulated within the traditional AI framework of building systems which rigidly constrain the processing of the system in a top-down, hierarchical, manner. These natural-language processors are manifested in the form of grammars which are decided, a priori, for processing the syntax, semantics, and pragmatics of natural-language utterances. One of the characteristics of traditional natural-language processing models, is that they are brittle due to their rigidity of processing. We argue that an approach to natural-language processing in the form of the Artificial Life (AL) paradigm will be more amenable to the flexible processing of utterances in a heterarchical manner. The AL approach has as a guiding principle the fact that the global behaviour of the system can emerge from the interactions of many components, each one following its own simple rules. Such processing will provide the capability of processing new

and dynamic forms of natural language. To date there have been little, or no, AL approaches to natural-language processing.

1 Introduction

It is commonly agreed that one of the most difficult problems in Artificial Intelligence (AI) is that of natural-language processing (see Partridge ([1])). There are many theories of how language can be processed by a computer program, some concentrating more on processing the structure, or syntax, of sentences (see Gazdar and Mellish ([2]), Pereira and Warren ([3]), and Woods ([4])), and others concentrating more on processing the meaning, or semantics, of utterances (see Schank ([5], [6]), Schank and Abelson ([7]), and Wilks ([8],[9],[10])). Recently, there has been an upsurge in research on the processing of pragmatics, or the usage of utterances (see Allen ([11], [12]), Ballim and Wilks ([13],[14]), Grosz and Sidner ([15]), Mc Kevitt ([16]), and Wilks and Mc Kevitt ([17])). However, all of these approaches treat language processing within the traditional paradigm of AI, where systems are designed in a framework such that the outcome of the system's processing is easily determined from its inputs.

In recent years a new paradigm for modelling intelligent behaviour has emerged, called Artificial Life (AL) (see Langton ([18])). This approach models intelligence from the point of view of exhibiting behaviour characteristic of natural living systems. Langton ([18], p. 2) says, "Artificial Life starts at the bottom, viewing an organism as a large population of *simple* machines, and works upwards *synthetically* from there – constructing large aggregates of simple, rule-governed objects which interact with one another nonlinearly in the support of life-like, global dynamics. The "key" concept in AL is emergent behaviour. Natural life emerges out of the organized interactions of a great number of nonliving molecules, with no global controller responsible for the behaviour of every part (his italics)." We explore the utility of emergent computation techniques, within the AL approach, for natural-language processing, as opposed to the AI approach, contrasting the differences of each. A design for processing natural language using the AL approach is described. We intend to implement this design in Quintus Prolog, and it is the only AL approach to natural-language processing that we know of.

2 Artificial Life

The Artificial Life (AL) approach to modelling intelligence comes from the point of view of modelling organisms as large populations of simpler agents. The simpler agents are rule-governed and interact with each other non-linearly. A key concept is *emergent behaviour*. The behaviour of a complete system is an emergence from the interactions of each individual agent, each following its own simple rules, in an organised way with other agents, where there is no global controller responsible for the behaviour of each agent.

For example, the behaviour of a flock of birds in flight would be modelled by determining the flocking rules of each individual bird (see Reynolds ([19])). There are no given rules for the behaviour of the flock as a whole. This global behaviour emerges from the activities and interactions of the individual birds. The AL approach has an advantage over explicit rule systems in that it is tolerant to variations in conditions which might not be foreseen. Another example would be modelling of a colony of ants. In this case one might provide specifications for the behavioural mechanisms of different *castes* of ants, and create lots of instances of each caste. The population of "automata" would be started from some initial configuration within a simulated two-dimensional environment. From that point on, the behaviour of the system would depend on the collective results of all of the local interactions between individual automata and between individual automata and features of their environment. There would be no single "dictator" automaton choreographing the ongoing dynamics according to some set of high-level rules for behaviour of the colony. The behaviour of the colony of automata would emerge out of the behaviours of the individual automata themselves, like in a real ant-colony (see Langton ([18]), p. 4).

The primary methodological approach of AL is one that models bottom-up, distributed and local behaviour. The approach can be modelled in a computer program by focusing on ongoing dynamic behaviour rather than on final results. The central features of computer-based AL models are:

- They consist of populations of simple programs or specifications.
- There is no single program that directs all of the other programs.
- Each program details the way in which a simple entity reacts to local situations in its environment, including encounters with other entities.
- There are *no* rules in the system that dictate global behaviour.
- Any behaviour at levels higher than the individual programs is therefore emergent.

AI is concerned with the generation of intelligent behaviour that bears no relationship to the method by which intelligence is generated in natural systems. The difference between AL and AI is that AL is concerned with the ongoing dynamics, rather than the state ultimately reached by the dynamics. AL researchers are not interested in building systems that reach any particular, a priori, designated solution.

3 Natural language processing

The traditional approach to natural-language processing in AI has been to use rules, or grammars, to dictate the global behaviour of a system which analyses incoming natural-language sentences. Many of the approaches use grammars of English to parse sentences into structures called *parse trees.* An example parse tree is shown in Figure 1.

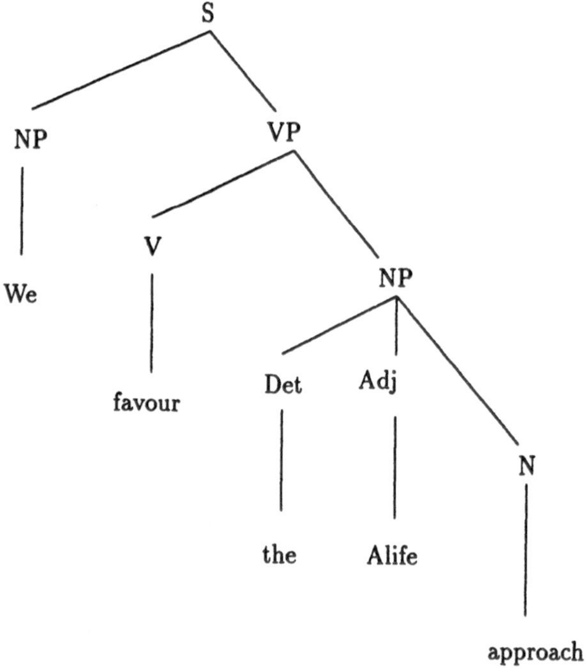

Figure 1: Sample parse tree

This parse tree represents the structure into which a traditional AI parser would parse the sentence, "We favour the Alife approach." The parser might use a grammar like that shown below, although, most likely, in a much more elaborate form. This grammar would be used to discover that the sentence consisted of the NP, "We", and the VP, "favour the Alife approach." Then, the VP would be broken down into V, "favour", and NP, "the Alife approach." Finally, the latter NP would be broken down into the Det, "the," the Adj ,"Alife," and the N, "approach."

S -- > NP VP

NP -- > Det Adj N

NP -- > Det Adj N PP

VP -- > V

VP -- > V NP

VP -- > V PP

VP -- > V NP PP

VP -- > V NP VP

PP -- > P NP

Much of the work in traditional computational linguistics and AI approaches to natural-language processing has used grammars of natural languages, such as English, to parse sentences into structures such as that shown in Figure 1. These structures are then augmented with various types of semantic processing. In fact, much of the work on semantic processing has also emphasised the primacy of semantics over syntax (see Wilks in [8], [9], [10] and Schank in [5], [6]).

One of the most difficult problems relating to semantics in natural-language processing is that of determining the correct sense of a lexically ambiguous word in context. For example, in the sentence, "The waiter served the lasagne," it is important that the system obtains the restaurant sense of serve, rather than the tennis-court one. Wilks' Preference Semantics system (see [8],[9] and [10]) was the first natural-language processing system to be explicitly designed around the need for lexical disambiguation. Wilks' system contained selectional restrictions, expressed in the form of templates. Restrictions were not fixed, but expressed as *preferences* within the system. A word that satisfied a preference was preferred, but if a word did not fit, the system would take the word that gave the best possible choice. Hence, the system always produced a solution. This enabled the handling of figurative usages of words, or metaphors, like in the example, "My car drank petrol." In the Preference Semantics system a selectional restriction on the verb "drink" would state that only animate entities can drink. However, the system would accept the sentence, forcing the knowledge structure for *car* to state that cars are able to drink. The knowledge structure for *car* would contain information about the fact that cars use gasoline, which is a liquid, and then infer that cars USing gasoline, are similar to cars DRINKing gasoline. All parts of speech were labelled with their respective preferences. For example, the adjective *big* was expected to qualify a physical object. The approach is similar to the predictive approaches of Riesbeck ([20], [21]) Schank et al. ([22]), and Riesbeck and Schank ([23]). For example in Riesbeck's analyser, a verb like *drink* would predict that the next object in an utterance would be a liquid.

The emphasis of work in natural-language processing has been in the processing of syntax and semantics using techniques similar to those just described. Some approaches emphasise the processing of syntax more than semantics, while others emphasise semantics more than syntax. Others balance the amount of syntax and semantics processing. While there has been much work on the processing of syntax and semantics in natural-language processing, there has been an upsurge recently on research into the processing of the use of language, or pragmatics. Original research in this area includes that of Schank and Abelson's ([7]) work on the use of world knowledge, and models of the beliefs, plans and goals of participants in a discourse for processing utterances. Recently, there has been work by Grosz and Sidner ([15]) on processing dialogues from the point of view of the structure of the dialogue. Allen ([11],[12]) provides a theory of processing natural language based on the mechanisms of planning and Ballim and Wilks ([13],[14]) provide a theory and computational model of how to model participants beliefs in discourse. Whether the traditional AI approaches to natural-language processing treat the modelling of syntax, semantics and pragmatics to an equal degree or not, they all have one thing in common: all of the approaches decide, a priori, using explicit rules, the global behaviour of the system. All the approaches are concerned with the state ultimately reached by the dynamics of the system.

4 Artificial life and natural language processing

Now, we shall consider how the AL technique can be applied to understanding natural-language sentences. Consider the sentence "John drinks water." Traditionally, to process this sentence, a parser would integrate syntactic processing in the form of grammar rules, and semantic processing in the form of semantic restriction rules. Also, there would be a lexicon indicating the parts-of-speech of each of the words, i.e. whether they are nouns or verbs, and possibly other information about the morphology of the words in different tenses, or declensions. A more complex lexicon would incorporate information about different senses of the words in different contexts.

Turning to the AL approach, imagine the individual words of a sentence as being low-level agents which have their own rules of behaviour[1]. These rules provide three types of information: syntactic information on structural constraints, semantic information on meaning constraints, and pragmatic information on usage constraints. For example, if we take the word, *drink*, as an agent, then it could contain the internal structure shown in Figure 2.

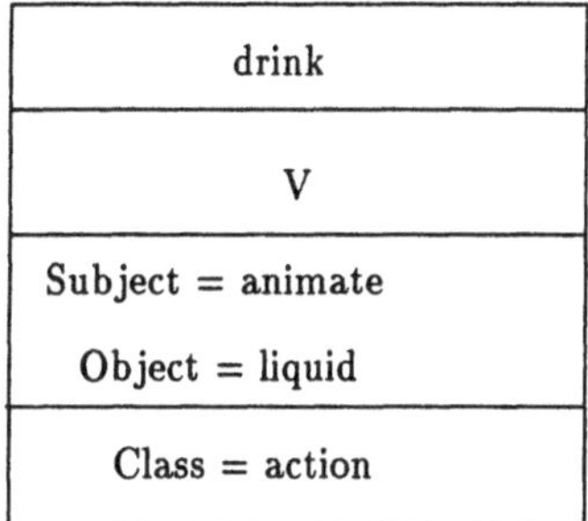

Figure 2: Agent structure for "drink".

The agent structure for *drink* has three boxes of information denoting information on syntax, semantics and pragmatics, from top to bottom respectively. First, the syntax box indicates that drink is a verb, V. Next, the semantics box indicates that *drink* prefers an *animate* subject, and the object of drinking to be a *liquid*. Finally,

[1]We are not interested here in morphological processing, or processing below the word level. Hence, we take words as atomic agents within the system.

the pragmatics box indicates that *drink* is of the class *action.* Also, we must take into account that *drink* can be a noun too. In this case the agent structure for drink will look as shown in Figure 3.

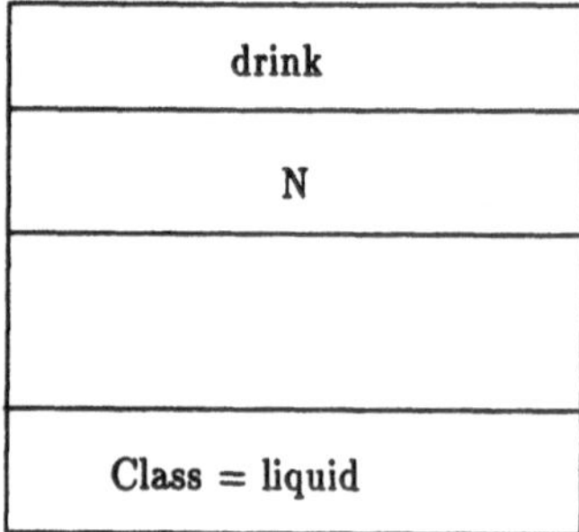

Figure 3: Agent structure for "drink" as noun.

A number of agents can be floating around in what we call the *agent pool* at any point in time. The agent pool could be thought of as a kind of *floating dictionary.* For example, the pool of candidate agents for the sentence, "John drinks water" is as shown in Figure 4. Note that although the pool includes different senses of the possible agents which occur in the sentence, "John drinks water," there are no connections between the agents.

The agents we have discussed so far are concerned with the modelling of words, just like a lexicon in the traditional approach to natural-language processing. We can call these agents, *word agents.* The pool should also contain *structure agents* for specifying structures that words can be integrated into. For example, structural agents might indicate that sentences are composed of noun phrases and verb phrases, noun phrases are composed of nouns, and verb phrases are composed of verbs and other noun phrases, as shown in Figure 5.

A question we must answer is: how do the agents combine to build a representation of a candidate sentence? The process works as follows. Before processing begins, the agent pool contains a number of word and structure agents which already exist as data. Next, words are entered as a stream, word by word, into the existing pool. There is no requirement that words be entered in the order they occur in a "normal" natural language sentence. The process is completed when the word stream ends in a terminator such as ".". Say, we drop the word *drink* into the pool.

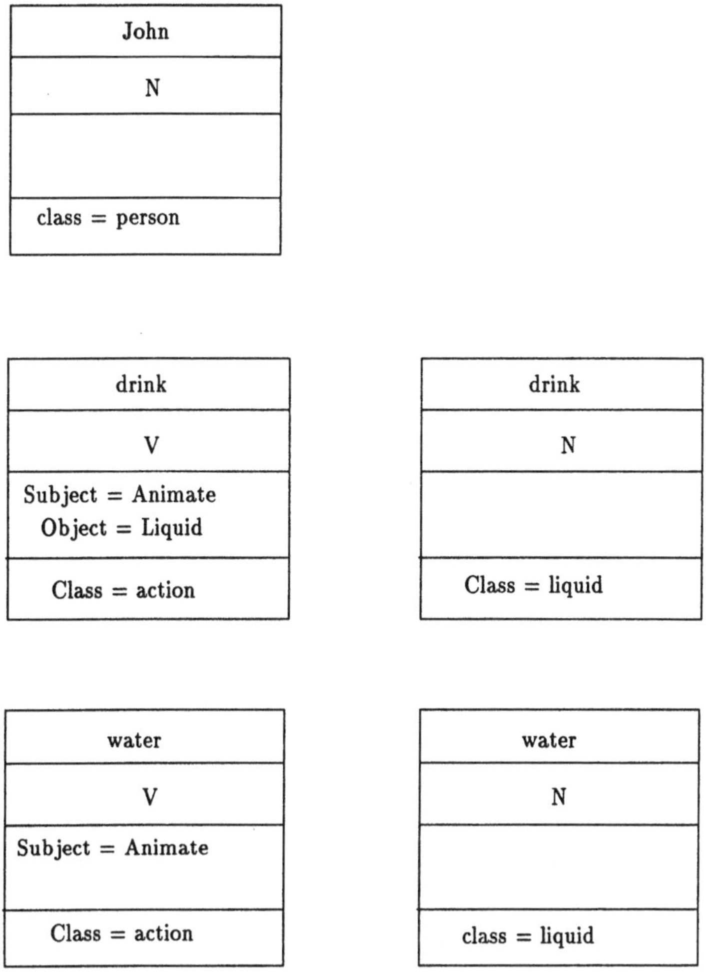

John
N
class = person

drink
V
Subject = Animate Object = Liquid
Class = action

drink
N
Class = liquid

water
V
Subject = Animate
Class = action

water
N
class = liquid

Figure 4: Sample agent pool

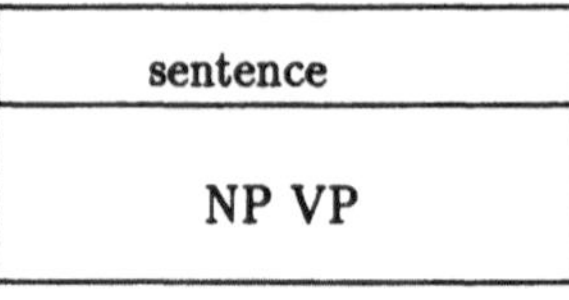

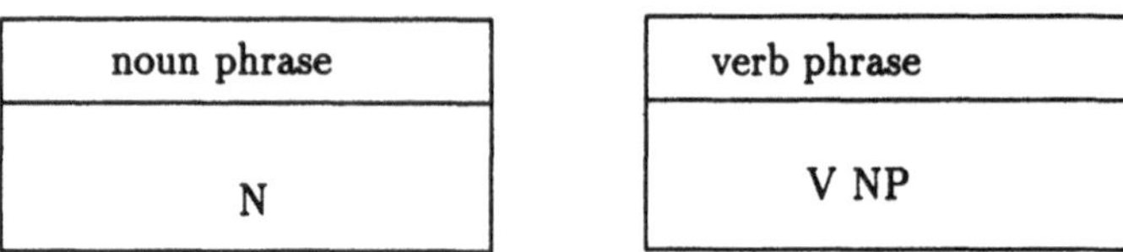

Figure 5: Sample structural agents.

Both the noun (N), and verb senses (V) for drink will be activated. However, there will be no *coagulation* of agents as yet, as only one word has been entered into the pool. Next, we can drop the word *John* into the pool. Now, the system will try and join *John* to one of the agents for *drink*, or both, if they are both appropriate. The structure and word agents check each other for possible linkage. The structure agent for the sentence agent discovers that drink can be a verb (V) or a noun (N), and also that John is a noun (N). However, this structure agent will rule out the *noun* agent for *drink* as a noun (N) cannot be followed by a noun (N). Next, the sentence agent checks information at the semantic level. It notices that the *verb* agent for *drink* asks for an animate subject, and that John is of class *person*, from the pragmatic information for John. It then checks the *person* agent and notes that people are animate. Also, the word agents *drink* and *John* check each other for suitability. As *drink* prefers liquids as objects, *John* must be the subject. Hence the pool stabilises as shown in Figure 6. The system always tries to find a match, just like Wilks' Preference Semantics (see [8], [9], [10]) system.

Next, the word water is added to the pool. The agent for *water* cannot be added to the pool, as it stands, as the wrong verb phrase structure agent has been selected. The system needs a verb phrase which includes an NP, and the noun (N) sense of water is selected. Hence, the current structure breaks down, and a new structure coagulates as shown in Figure 7. This is the final structure for the sentence, "John

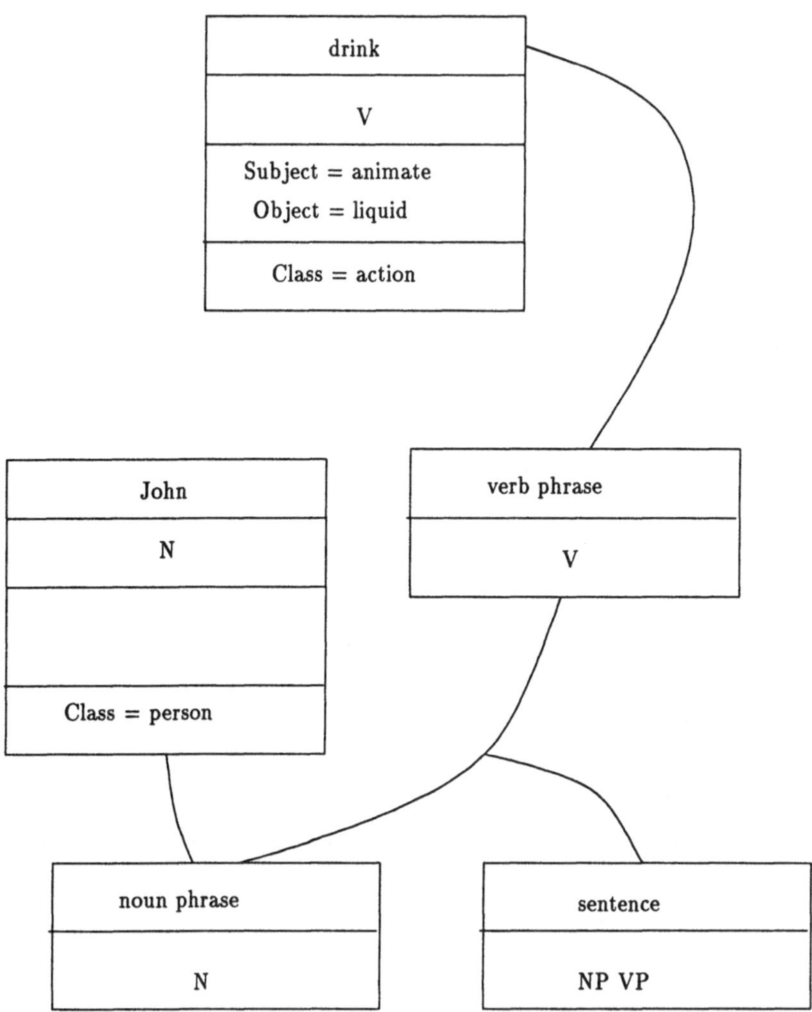

Figure 6: Agent pool for "John drinks"

drinks water."

You may now argue that, so far, the process we have described, is not, in any way different from the traditional AI approach to natural-language processing. However, there is one fundamental difference. In the traditional AI approach there is a notion of global control, where the system is trying to obtain a parse of an sentence, with respect to some specific grammar rule. There is a definite goal towards which the word sequences are directed. However, in the AL approach we do not know the result of a parse, or in what direction the parsing will go, until we see the result produced by the system.

Note that the approach demonstrated here has the words entered into the word pool in their natural left-to-right order. There is no reason as to why this should necessarily be the correct way of loading the word stream into the pool. Right-to-left could be just as good, or we could have started in the middle, or initially segment the input stream with a first pass as Wilks does in [8]. The important point, is that the more input we place in the pool, the more constraining the context is for further words coming in. Hence, early words and sentences provide a context which assists with any unnecessary disambiguation, in a similar fashion to the PDP approach described in McClelland and Kawamoto in [24].

As well as combining together, structures will often split apart. This will happen when a constraint is violated. For example, in the sentence "John holds the ball and drinks the water", it could happen that "the ball" is considered as a subject for "drinks" by a syntactic agent. However, the semantic constraints will try to break up this combination, since "drinks" will seek an *animate* agent as its subject. So far we have considered sentences which do not involve figurative usage. Let's now move on to look at how such sentences would be processed.

5 Metaphor processing

One of the most prevalent problems in natural-language processing has been that of metaphor. Consider the sentence, "The car drinks petrol." The agent pool will be satisfied with "The car" and will have no problem building a structure for it. However, when "drinks" is added to the agent pool there will be a problem. There will be a semantic constraint violation since the pragmatic information for the *car* agent indicates that they are machines, and it is easy to derive that machines are inanimate. Hence, "car" would not be allowed as the subject of the *drink* agent. The pool would then contain a linking of the agents for "the" and "car" but the agent for *drink* would remain on its own. Next, the word *petrol* is added to the pool. This would link together with drinks as the pool would notice that petrol is a liquid

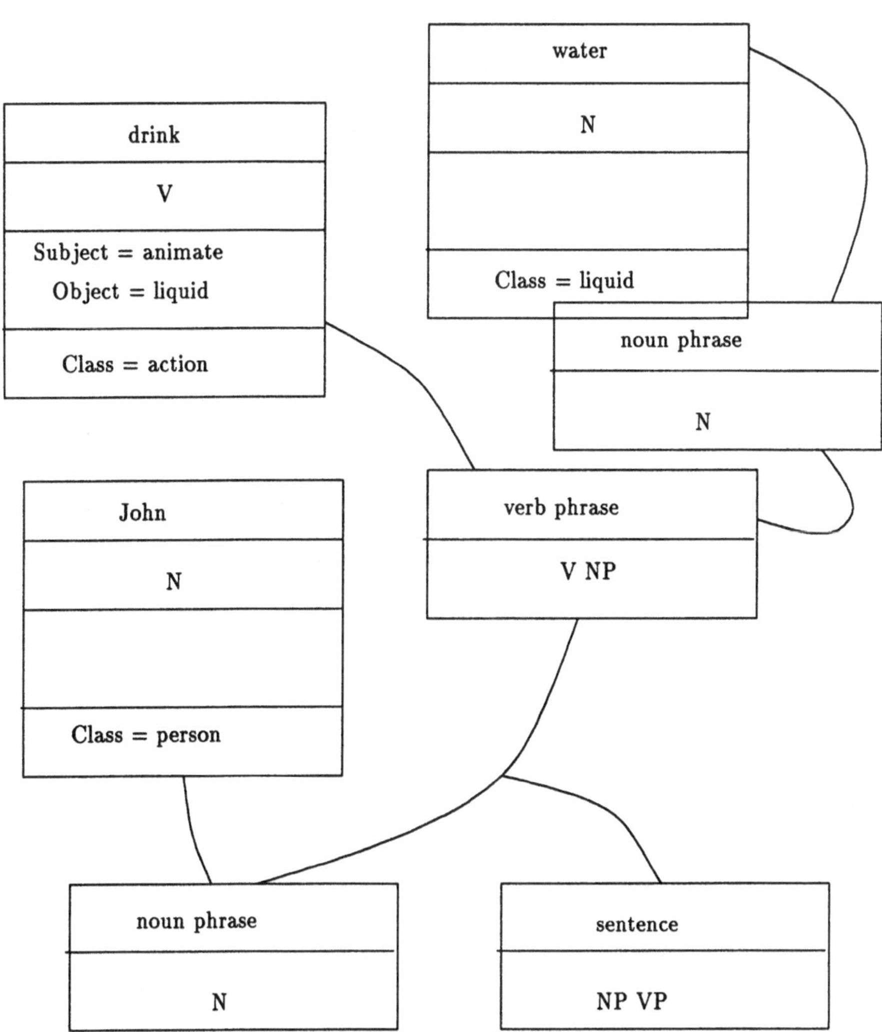

Figure 7: Agent pool for "John drinks water"

from the *liquid* agent, and liquids are drunk from the *drink* agent. Hence, the pool will now have a link between "the" and "car" and between "drinks" and "petrol." However, the *sentence* agent will still not be able to determine a link between the agents for car and drinks. Yet, the system will have noticed now that there is a terminator in the input. It will then try to force a link between the *car* and *drink* agents. The semantic constraint that cars are machines will be relaxed. A final structure will be formed with "the car" as subject noun phrase. However such a structure would have a lower degree of satisfaction than say, "John drinks water." The system could incorporate a model of satisfaction, or tolerance, which measures the degree of constraint satisfaction found (see Hofstadter ([25])).

The system allows both the relaxation of syntactic and semantic constraints. Relaxing a semantic constraint gives rise to the processing of *metaphors.* In the example just described there is a metaphorical interpretation of the agent *car* as an *animate* object. This could be achieved by allowing the pragmatic constraints of objects to be flexible. Hence, agents such as *car* could be updated so that their pragmatic information contains the fact that they can be animate. We show this promotion of pragmatic information for the agent *car* from being a *machine* to being *animate* in Figure 8 below.

The promotion could be done by hand by the programmer updating the word agent for car. However, it would be simple to modify the system so that it automatically updated the pragmatic component of agents automatically from new input. Hence, the system's grasp of language and conceptual organisation would evolve with experience. Thus metaphor becomes the basis of language understanding and development, rather than being seen as a quirk in an otherwise cleanly defined language. This seems to be a much more helpful approach to the role of metaphor in language, and one that is stressed in Lakoff and Johnson in [26]. The problem of how modification of agents can be done effectively, or how learning can take place in emergent systems, is described in detail by Rowe in [25].

Just as we have shown how pragmatic information can be updated within agents, syntactic information can be updated as well. The syntactic component of an agent description could be augmented to make the system tolerant to grammatically ill-formed input. Hence, we would then have a system which would be able to parse ill-formed human dialogues, texts and even new syntactic forms[2]. The system would assemble the best interpretation of incoming sentences given the constraints, instead

---

[2]For example, many new syntactic forms exist in the novel "Ulysses" by James Joyce (see Joyce [28]). Most current natural language systems would have great difficulty in trying to parse the non-conventional sentences in Ulysses. However, it is important to note that a morphological component would need to be added to our agent structures to deal with Joyce's work.

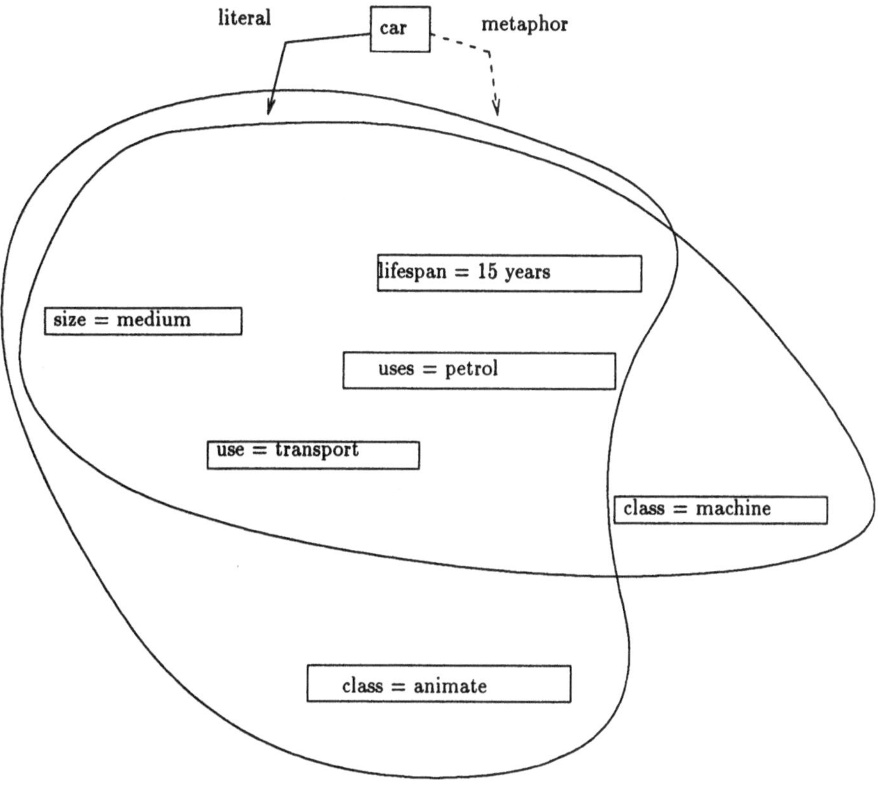

Figure 8: Promotion of pragmatic information for 'car'

of merely reporting a syntactic, semantic or pragmatic error, or a message saying the input cannot be processed. As with the metaphor case just described, the approach to syntactic tolerance would enable a grammar to be viewed not as a given set of fixed rules, but something that emerges and evolves over time, according to the system's experience with parsing sentences.

One of the major applications of theoretical work in natural-language processing is machine translation where algorithms are developed to translate utterances from one language into another. Considering our AL model words from different languages would easily point to the same (or similar) groups of semantic constraints. Much of the time, words in one language do not correspond exactly with those of another. For example, the the German word *wissen* is some kind of subset of the English *know*. This could be represented by using the promotion technique again. The structure would be represented as shown in Figure 9 below.

We can see that he German word *wissen* is analogous to the English word *know* with some exceptions. This process could be pursued further for an application of the AL technique to machine translation.

6 Conclusion

It is concluded here that the AL approach to modelling intelligent behaviour is useful for natural-language processing. The approach enables the bottom-up parsing of sentences into syntactic and semantic structures. The approach does not bias the system into trying to force one parse over another, but allows the system to determine the best parse it can find.

There are many similarities between the AL approach and connectionist approaches to natural-language processing such as those described in McClelland and Kawamoto ([24]), Sharkey et al. ([29]), and Sharkey and Sharkey ([30]). However, one major difference is the fact that one can determine exactly the path that the AL model takes during processing, as it is at the symbolic level. Hence, the AL approach has both the bottom-up freedom of connectionist systems and the transparency of symbolic systems.

The AL approach gives us several advantages, including tolerance of bad grammar, ill-formed input and metaphor understanding. The model also enables us to explore the learning and development of language and new ideas concerning machine translation.

We are currently implementing the AL model in Prolog and future work will involve comparing simulations of the AL model with simulations of traditional models. Although the AL approach to intelligence is growing rapidly, there are to our

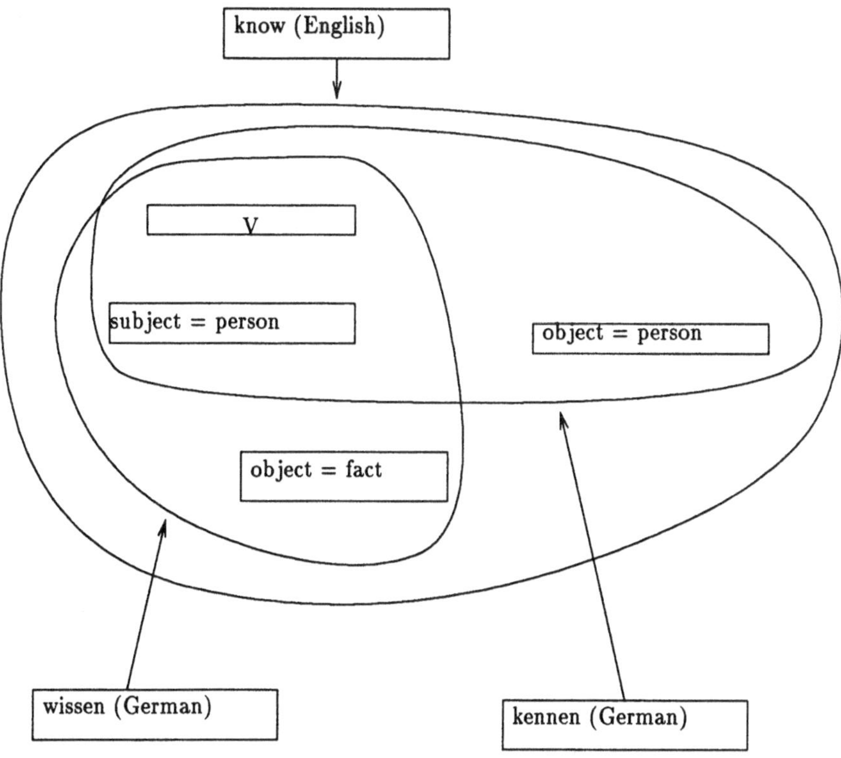

Figure 9: Representation for *wissen*

knowledge no other AL models of natural language processing. The closest model is that proposed in Small ([31]) and Small and Reiger ([32]).

7 Acknowledgements

We would like to thank Conor Doherty, Mark Ennis, Gerard Hartnett, Noel Murphy, Barry McMullin and Ronan Reilly for providing comments on this work. Paul Mc Kevitt would like to thank the School of Electronic Engineering at Dublin City University for support as a Visiting Fellow while conducting this research.

8 References

1. Partridge, Derek. *A new guide to Artificial Intelligence.* Norwood, New Jersey: Ablex Publishing Corporation, 1991.
2. Gazdar, Gerald and Chris Mellish. *Natural language processing in PROLOG: an introduction to computational linguistics.* Workingham, United Kingdom: Addison-Wesley, 1989.
3. Pereira, Fernando C. and David Warren. *Definite clause grammars for language analysis: a survey of the formalism and a comparison with augmented transition networks.* In Artificial Intelligence, 13, 3, 231-278, 1980.
4. Woods, W.A.. *Transition network grammars for natural-language analysis.* In Communications of the ACM, 13(10), 591-606, October, 1970.
5. Schank, Roger C.. *Identification and conceptualizations underlying natural language.* In Computer Models of Thought and Language, R. Schank and K. Kolby (Eds.), San Francisco, CA: Wh Freeman and Co., 1973.
6. Schank, Roger C.. *Conceptual information processing.* Fundamental Studies in Computer Science, 3, Amsterdam: North-Holland, 1975.
7. Schank, Roger C. and Robert P. Abelson. *Scripts, plans, goals and understanding: an inquiry into human knowledge structures.* Hillsdale, NJ: Lawrence Erlbaum Associates, 1977.
8. Wilks, Yorick. *Preference semantics.* In Formal semantics of natural language, Keenan, Edward (Ed.), Cambridge, United Kingdom: Cambridge University Press, 1975. Also as, Memo AIM-206, Artificial Intelligence Laboratory, Stanford University, Stanford, California, USA, July, 1973.
9. Wilks, Yorick. *An intelligent analyzer and understander of English.* In Communications of the ACM, 18(5), 264-274, May, 1975. Also in, *Readings in Natural Language Processing,* Barbara Grosz, Karen Sparck Jones and Bonnie Webber (Eds.), 193-204, Los Altos, California: Morgan Kaufmann, 1986.

10. Wilks, Yorick. *A preferential, pattern-seeking semantics for natural language inference.* In Artificial Intelligence, 6, 53-74, 1975.

11. Allen, James F.. *A plan-based approach to speech act recognition.* Doctoral Dissertation. Also as, Technical Report No. 131/79, University of Toronto, Toronto, Canada, February, 1979.

12. Allen, James F.. *Recognising intentions from natural language utterances.* In Computational Models of Discourse, M. Brady and R.C. Berwick (Eds.), 107-166, Cambridge, MA: MIT Press, 1983.

13. Ballim, Afzal and Yorick Wilks. *Stereotypical belief and dynamic agent modeling.* In Proceedings of the Second International Conference on User Modelling, University of Hawaii at Manoa, Honolulu, Hawaii, USA, 1990.

14. Ballim, Afzal and Yorick Wilks. *Artificial Believers.* Hillsdale, New Jersey: Lawrence Erlbaum Associates, 1991.

15. Grosz, Barbara and Candy Sidner. *Plans for discourse.* In Intentions in Communication, Cohen, P.R., J.L. Morgan, and M.E. Pollack (Eds.), Cambridge, MA: Bradford Books, MIT Press, 1990.

16. Mc Kevitt, Paul. *Analysing coherence of intention in natural language dialogue.* Ph.D. Thesis, Department of Computer Science, University of Exeter, GB- EX4 4PT, Exeter, E.C., 1991.

17. Wilks, Yorick and Paul Mc Kevitt (Eds.). *Proceedings of the Fifth Rocky Mountain Conference on Artificial Intelligence (RMCAI-90), Subtitled, 'Pragmatics in Artificial Intelligence'.* Computing Research Laboratory, Dept. 3CRL, Box 30001, New Mexico State University, Las Cruces, NM 88003-0001, USA, June, 1990.

18. Langton, Chris. *Artificial Life.* In C. G. Langton, (Ed.), 1-47, Artificial Life, Addison-Wesley Publishing Company, 1989.

19. Reynolds, C.W.. *Flocks, herds and schools: a distributed behavioural model.* In Computer Graphics, 21, 4, 25-34, 1987.

20. Riesbeck, Chris. *Computational understanding: analysis of sentences and context.* Doctoral dissertation, Computer Science Department, Stanford University, May, 1974.

21. Riesbeck, Chris. *Conceptual analysis.* In Conceptual information processing (Fundamental studies in computer science 3), Roger Schank (Ed.), 83-156, Amsterdam: North-Holland, 1975.

22. Schank, Roger C., Neil Goldman, Charles Reiger, and Chris Riesbeck. *Inference and paraphrase by computer.* In Journal of the Association of Computing Machinery, 22(3), July, 309-328, 1975.

23. Riesbeck, Chris, and Roger C. Schank. *Comprehension by computer: expectation-*

based analysis of sentences in context. In Studies in the perception of language, Willem J. Levelt and Giovanni B d'Arcais (Eds.), New York: John Wiley, 1978. Also as Research Report 78, Department of Computer Science, Yale University, October, 1976.

24. McClelland, J.L. and A. H. Kawamoto. *Mechanisms of sentence processing.* In Parallel Distributed Processing, J. L. McClelland and D. E. Rumelhart (Eds.), 19, 272-325, Cambridge, Mass.: MIT Press, 1986.
25. Hofstadter, Douglas R.. *The architecture of Jumbo.* In Proceedings of the International Machine Learning Workshop, R. S. Michalski (Ed.), University of Illinois at Urbana-Champaign, 1983.
26. Lakoff, George and M. Johnson. *Metaphors we live by.* Chicago, Illinois: University of Chicago Press, 1980.
27. Rowe, J. E.. *Emergent creativity: a computational study.* Ph.D. thesis, Department of Computer Science, University of Exeter, GB- Exeter EX4 4PT, E.C., 1991.
28. Joyce, James. *Ulysses.* New York: Random House, 1961.
29. Sharkey, Noel E., R.F.E. Sutcliffe, and W.R. Wobcke. *Mixing binary and continuous connection scheme: for knowledge access.* In Proceedings of the Fifth American National Conference on Artificial Intelligence (AAAI-86), Philadelphia, PA, USA, August, 1986.
30. Sharkey, Noel E. and Amanda J.C. Sharkey. *KAN: A knowledge access network model.* In Communication failure in dialogue and discourse: detection and repair processes, Ronan Reilly (Ed.), , Amsterdam: Elsevier Science Publishers/North-Holland, 1987.
31. Small, S.. *Daemon timeouts: limiting the life spans of spontaneous computations in cognitive models.* In Proceedings of the Third Annual Meeting of the Cognitive Science Society, Berkeley, California, USA, 1981.
32. Small, S. and Charles Reiger. *Parsing and comprehension with word experts: a theory and its realisation.* In Strategies for Natural Language Processing, M.D. Ringle and W. Lehnert (Eds.), Hillsdale, New Jersey: Lawrence Erlbaum Associates, 1982.

Section 5:

Deduction

Who is Telling the Truth......
Cognitive Processes in Meta-Deductive Reasoning

Ruth M.J. Byrne[1], **P.N. Johnson-Laird**[2], **and Simon J. Handley**[3]

Abstract

How can the mechanisms underlying both human and computer reasoning develop to deal with novel or complex inferences ? The answer may lie in research in the cognitive science of *meta-deductions*. Some meta-deductions, such as the inferences we make when we try to think about what other people are thinking, are central to successful social and professional interactions, and to the interactions of people with computers. Other meta-deductions, such as the inferences we make when we try to work out the truth or falsity of alternative states of affairs, can seem both novel and complex. Consider, for example, an island inhabited by two sorts of people, knights who always tell the truth, and knaves who always lie. Suppose you overhear a conversation between two of these individuals. A asserts, *I am a knight and B is a knight.* B asserts, *A is a knave.* Can you work out whether A is a knight or a knave, and whether B is a knight or a knave? In this chapter, we examine experimental data on the sorts of inferences that human reasoners make from these puzzles. The data help us to chose from among the programs developed to simulate theories of the mechanisms underlying these inferences.

1 Inferences about truth

When passers-by look curiously at you while you're waiting on a street corner to meet someone, do you glance conspicuously at your watch? You probably don't want to know the time. You think that the passers-by think that you're suspicious, and you want to make them think otherwise. Your action is based on your ability to make inferences about other people's inferences. Some people seem to be better than others at making such *meta-inferences* -- for example, successful politicians or business tycoons seem to be skilled at "outwitting" their opponents. The study of the development of meta-inferential skill has emerged only recently in cognitive psychology [1, 2, 3, 4] and its findings have important implications for the construction of artificially intelligent reasoning programs.

---

[1] Department of Psychology, University of Dublin, Trinity College, IRELAND.
[2] Psychology Department, Princeton University, USA.
[3] School of Psychology, University of Wales College of Cardiff, BRITAIN.

1.1 Deduction

The study of deduction in people and in computers has been concerned primarily with mechanisms for making valid inferences, that is, mechanisms for the construction of conclusions that must be true if the premises on which they are based are true [for a review see 5]. Suppose a reasoner knows the following information:

> If a patient tells the casualty doctor that he or she was bitten by a dog,
> then the doctor will give him or her a tetanus injection.

When the reasoner is told the further information:

> Jane told the casualty doctor that she was bitten by a dog.

he or she can readily make the valid deduction:

> Therefore, the casualty doctor gave her a tetanus injection.

Individuals are largely unaware of the processes by which they make such inferences. They can introspect on the input to these processes -- the premises, and on the output -- the conclusion. They might be aware of the intermediate outputs between the input and output -- for example, the intermediate conclusion that Jane is a patient. But, they are unlikely to have any conscious access to the mental machinery that underlies the inference. To gain insights into the workings of this machinery requires more than individual reflection. Cognitive scientists have developed alternative theories of the nature of this machinery, and they have constructed computational models to simulate their theories. We will describe these theories in the second section of this chapter. To test which theory is closest to the way human reasoners make inferences, cognitive scientists have compared the predictions that each theory makes about reasoning, to the experimental evidence on human reasoning. We will describe one such test in the third section. Our descriptions will centre on the domain of meta-deduction, to which we now turn.

1.2 Meta-deduction

A recent development in the study of reasoning is the attempt to understand how reasoning skill develops in adults [1, 2, 3]. To study these processes we have adopted a particular sort of puzzle, made popular by the philosopher Smullyan [6, see also

7, 8]. Reasoners find these puzzles very difficult, but they appear to develop strategies to solve them, swiftly and spontaneously. The puzzles are of the following sort:

> *Imagine an island inhabited by two sorts of people, knights who always tell the truth, and knaves who always lie. Suppose you overhear a conversation between two individuals.*
> *A asserts: I am a knight and B is a knight.*
> *B asserts: A is a knave.*
> *Your task is to decide on the basis of this information, if possible, whether A is a knight or a knave, and whether B is a knight or a knave.*

Two alternative theories of the cognitive processes underlying these inferences have emerged and we will outline each of them in the next section.

2 Two reasoning algorithms

2.1 A hypothetico-deductive algorithm based on syntactic rules

One theory of reasoning, developed by Lance Rips, is based on formal, syntactic rules of inference [1, 3]. Rips's program, written in Prolog and called **ANDS** (for **A** **N**atural **D**eduction **S**ystem [9]), takes as input premises of the form:

> A asserts: I am a knight and B is a knight.
> B asserts: A is a knave.

and translates them to abstract representations of the form:

1. A asserts A and B
2. B asserts not-A

using content-specific rules of inference based on the meaning of knight and knave. It contains content-independent rules of inference based on the logical constants, *or*, *and*, *not*, and *if*. These rules can be used in the construction of proofs of conclusions. The proofs are constructed in accordance with an algorithm based on the following sort of *hypothetico-deductive* strategy:

- *Hypothesise the first assertor is telling the truth.*
    - *Follow up the consequences of this hypothesis, using inference rules to derive conclusions.*
    - *If it implies the second assertor is telling the truth or telling a lie, follow up its consequences, and so on.*
    - *If a contradiction is reached, reject the hypothesis.*
- *Hypothesise the first assertor is telling a lie; do likewise.*
- *If a consistent assignment is reached for an assertor, the assertors status is established, otherwise, it is undetermined.*

When this algorithm is applied to the premises of the inference, the program begins by assuming that A is telling the truth:

3. A *[hypothesis-creating rule]*

and infers that since A is a knight, A is telling the truth, and hence A's assertion is true:

4. A and B *[rule for meaning of knight, from step 3 and premise 1]*

It then infers that B is a knight since A says so:

5. B *[conjunction elimination rule, from step 4]*

It follows up the consequences of this inference, and infers that B's assertion is true:

6. not-A *[rule for meaning of knight, from step 5 and premise 2]*

It detects the contradiction that results from the sequence of inferences by joining the two pieces of information it has about A, that A is a knight, and that A is a knave:

7. A and not-A *[conjunction introduction rule, from steps 3 and 6]*

The hypothesis that A is telling the truth has led to this contradiction. According to the principle that any hypothesis that leads to a contradiction is false, it rejects its initial hypothesis. It concludes that A is telling a lie:

8. not-A *[from contradiction in 7 and hypothesis in 3]*

It now proceeds to follow up the consequences of the conclusion that A is a knave. It infers that A's assertion is false and negates it:

9.	not (A and B)	*[rule for meaning of knave, from steps 8 and 1]*

and it unpacks this negation to an equivalent one:

10.	not-A or not-B	*[DeMorgans law, from step 9]*

It has already deduced that not-A is the case. But, it has not yet established whether not-B is the case. It assumes not-B, and B, in turn, to examine the consequences of these assumptions. From the hypothesis that not-B is the case, it once again reaches a contradiction:

11.	not-B	*[hypothesis-creating rule]*
12.	A	*[rule for meaning of knave, from step 11 and premise 2]*
13.	not-A and A	*[conjunction introduction rule, from steps 12 and 8]*

and so it rejects the hypothesis that not-B is the case. It concludes that B is the case:

14.	B	*[from contradiction in 13 and hypothesis in 11]*

The final step is to follow up the consequences of the conclusion that B is the case:

15.	not-A	*[rule for meaning of knight, from step 14 and premise 2]*

The derivation has led to a consistent assignment: not-A is the case, and B is the case. The program deduces that A is a knave and B is a knight.

In summary, the program contains syntactic rules of inference, and it relies on a single, deterministic hypothetico-deductive strategy.

2.2 A simple-strategies algorithm based on a semantic procedure for the construction of models

Our view of the cognitive processes underlying meta-deductions is that they depend on the manipulation of mental models [10, 5]. Our program, written in Lisp and called **META-PROPAI** (an **AI** program for **meta-prop**ositional reasoning [2]) makes meta-deductions by first constructing a set of models corresponding to the premises. We will present some examples of the models the program constructs, before we describe its simple-strategies algorithm.

A simple conjunction, such as:

A is a knight and B is a knight.

is represented by tokens corresponding to each proposition, which are inserted into a single model (a single list in the program):

A B

The negation of a conjunction:

It is false that A is a knight and B is a knight.

is represented in a set of models:

A ¬B
¬A B
¬A ¬B

Each line represents a separate model which corresponds to a state of affairs in the world (a separate list in the program), and the "¬" symbol represents negation [see 11, 12]. In the first model, A is a knight and B is not, in the second model B is a knight and A is not, and in the final model, neither A nor B are knights.

The program can represent an assertor's assertion, such as:

A asserts A and B.

It constructs a model or set of models of the assertion, and it includes a token corresponding to the assertor. If the assertor is telling the truth, then the models contain a token corresponding to the assertor, and the set of models corresponding to the assertor's assertion:

A: A B

If the assertor is lying, then the models contain a token corresponding to the negation of the assertor, and the set of models corresponding to the negation of the assertor's assertion:

$\neg A$: $\neg A$ B

$\neg A$ $\neg B$

Notice that the set of models for the negated assertion does not include the model:

A $\neg B$

because this model contains the information that A is a truthteller, which conflicts with the information that A, the assertor, is a liar: $\neg A$.

The algorithm for making meta-deductions is based on many simple strategies. We will illustrate the nature of these strategies by describing just one of them, a *hypothesise-and-match* strategy, which has the following structure:

- *Hypothesise the first assertor is telling the truth, and construct a model of his or her assertion.*
  - *Follow up the consequences of this hypothesis, using a semantic procedure based on models to derive conclusions.*
  - *If a contradiction is reached about the status of the first assertor, (and so a definite assignment about the first assertor is possible), match this assignment to the second assertor's assertion.*
  - *Otherwise, fail.*

Consider the premises of our earlier example:

A asserts A and B.

B asserts not-A.

The hypothesise-and-match strategy requires the program to construct a model corresponding to the truth of the first assertor, and a model of this assertor's assertion:

A: A B

Since B is true in this model, the program constructs a model of B's assertion:

B: $\neg A$

When the model-combining procedures attempt to integrate these two models, they detect a contradiction, A and not-A, and so the initial hypothesis -- that A is the case -- is rejected. So far, the strategy seems equivalent to the hypothetico-deductive strategy of the rule-based program, with the only difference that it is based on models instead of rules of inference. But, from this point on, the two strategies differ. The model-based program detects that it has reached a definite assignment about A -- it has concluded that not-A is the case. And at this point it attempts to match its conclusion to the second assertor's assertion. B asserts that not-A is the case, which matches the conclusion that the program has reached. The program infers that B's assertion is true, and hence that B is telling the truth. It concludes that not-A is the case and that B is the case.

The strategy works only for those problems for which the program reaches a definite assignment for one assertor, and successfully matches, or mismatches, this assignment to another assertor's assertion. It is a limited strategy, but in the cases where it does work, it is an efficient short-cut. It eliminates the need to make many further inferences, and the need to hypothesise about the truth and falsity of the assertors.

In summary, the program contains a procedure for constructing models, and it relies on a host of simple, heuristic strategies.

3 A Test of Two Theories

Which of the two alternatives is a closer approximation to the way human reasoners make meta-deductions? We have tested the theories by deriving from them alternative predictions about human reasoning performance, and we have checked these predictions experimentally. To illustrate, we will describe just one of our experimental tests of the two theories. They make diverging predictions about the difficulty of meta-deductions. The model theory predicts that one meta-deductive problem will be easier than a second, if reasoners can use a simple strategy to solve the first but not the second. The rule theory makes no such prediction, because it proposes that reasoners use a single strategy to solve all problems. The model theory proposes that hypothesising -- in particular, formulating an appropriate hypothesis -- is a source of difficulty in meta-deductions (and so reasoners develop simple strategies, many of which are designed to circumvent the need to make hypotheses). The rule theory makes no such prediction, since it proposes that reasoners always begin by hypothesising that the first assertor is telling the truth.

3.1 An Experiment on Meta-Deductions

We tested two of the diverging predictions of the model theory and the rule theory. First, we gave reasoners meta-deductions which could be solved by the hypothesise-

and-match strategy and meta-deductions which could not be solved by this simple strategy. The model theory predicts that reasoners should find the first sort of problem easier than the second. The rule theory predicts that the two sorts of problems should be of equal difficulty. Second, we gave one group of reasoners some help with the construction of the initial hypothesis for each problem, and we gave a second group of reasoners no help. We gave them a ready-made hypothesis that the first assertor was a knight or that the first assertor was a knave. This hypothesis was either appropriate or inappropriate. The rule theory predicts that subjects should find the hypothesis that the first assertor is a knight more helpful than the hypothesis that the first assertor is a knave; the model theory predicts that subjects should find the appropriate hypothesis more helpful than the inappropriate one.

We constructed a set of 48 problems of the following sort:

A asserts: I am a knight and B is a knight.

B asserts: A is a knave.

The problems consisted of two premises and each contained an assertion made by A and B. A's assertion referred to the status of A and B as either a knight or a knave -- and was based on a conjunction, an exclusive disjunction, or an inclusive disjunction. B's assertion referred to A's status as a knight or a knave. These 4 x 3 x 2 possibilities resulted in twenty-four problems, each of which was presented twice [see 13 for details]. We gave the problems to twenty paid volunteers from the University of Wales College of Cardiff. Their task was to tick one of three boxes, labelled "knight", "knave" and "undetermined" for A, and one of three similar boxes for B.

Half of the problems with determinate answers could be solved by the simple hypothesise-and-match strategy, and half could not. The subjects made the correct inferences for the problems that could be solved by the simple strategy on 44% of occasions. They made the correct inferences for the problems that could not be solved by the simple strategy on only 30% of occasions. This difference is statistically reliable [$F(1, 18) = 7.89$, $p < 0.01$]. These results corroborate the model theory and go against the rule theory.

In the experiment, the twenty subjects were assigned to two groups. We gave one group problems such as the one above, and we gave the second group the same problems, but accompanied by a hypothesis:

Assume A is a knave.

A asserts: I am a knight and B is a knight.

B asserts: A is a knave.

For one version of the problems, the hypothesis was the appropriate one, that is, it corresponded to A's correct status, as in the example. For the second version of the problems, the hypothesis was the inappropriate one, that is, it did not correspond to A's correct status, e.g.:

> Assume A is a knight.
>
> A asserts: I am a knight and B is a knight.
>
> B asserts: A is a knave.

We gave the subjects in the second group both versions of the problems, presented at random. The subjects who were given a hypothesis did not make more correct inferences when they were given the hypothesis that A is a knight (49%) than when they were given the hypothesis that A is a knave [36%, $F(1, 18) = 3.64$, non-significant], contrary to the predictions of the rule theory. They made more correct inferences when they were given the appropriate hypothesis (52%) than when they were given the inappropriate hypothesis [33%, $F(1, 18) = 6.84$, $p < 0.01$], as the model theory predicts.

But, it is the interaction between the effects of the two primary variables that is most interesting. Table 1 presents the percentages of correct inferences made by the subjects given no hypothesis, given an appropriate hypothesis, and given an inappropriate hypothesis, to the problems that can be solved by a simple strategy and the problems that cannot be solved by a simple strategy.

Table 1: The percentages of correct inferences to the problems in the experiment

	Simple strategy	*No simple strategy*
No hypothesis	44	18
Appropriate hypothesis	46	58
Inappropriate hypothesis	41	25

The results show that helping subjects with their initial hypothesis has the greatest effect for problems that cannot be solved with a simple strategy. For these problems, the subjects who were given an appropriate hypothesis made reliably more correct inferences (58%) than the subjects who were not given a hypothesis [18%, $F(1, 72) =$

10.69, $p < 0.002$].[4] For the problems that could be solved by the simple strategy, the two groups of subjects made the same percentage of correct inferences (46% and 44%). The subjects who were given the appropriate hypothesis were not simply accepting it as true and failing to reason further: when they were given the inappropriate hypothesis, there was no decrease in the number of correct inferences they made (25%) compared to the group who received no hypothesis [18%, $F(1, 72) = 0.54$, non-significant]. These results support the proposal that reasoners develop simple strategies to help them overcome their difficulties with the construction of hypotheses.

4 Conclusions

Meta-deductions are hard when human reasoners do not have a simple strategy to solve them. The model-based algorithm relies on the operation of a host of simple strategies, and so it is consistent with this finding. Meta-deductions that cannot be solved by a simple strategy are hard because reasoners must construct and pursue the consequences of hypotheses about each individual. The results of the experiment support the model-based view of reasoning [see also 14, 15, 16, and 17].

Some reasoners may develop their meta-deductive skill to the point where they are capable of using a hypothetico-deductive strategy similar to the one embodied in Rips's program. We believe that this level of skill is beyond the reach of naive reasoners, because of the limitations in human processing capacities. Instead naive reasoners develop a host of simple strategies, or heuristic short-cuts, that allow them to extend their reasoning abilities to novel and complex problems. We have developed an algorithm based on simple strategies, guided by the results of our experiments with human reasoners. The experimental results suggest that human reasoners develop simple strategies according to clear principles, for example, they develop strategies that obviate the need to make hypotheses. A potentially fruitful line of enquiry for understanding the development of strategies -- in both human and computer reasoning mechanisms -- is to establish how components of different simple strategies may be combined to result in a new strategy. The emergence of new strategies from old ones seems to lie at the heart of the flexibility of human reasoning.

---

[4] The results for the no-hypothesis group are collapsed over the two (identical) versions of the problems.

Acknowledgements

The research reported in this paper is supported by grant no. *R000 23 2491,* to Ruth Byrne from the ***Economic and Social Research Council*** of Britain. We thank the Computer Science Department, University College Dublin, for facilities to conduct some of the research, and Mark Keane for helpful comments on an earlier version of the chapter.

References

[1] Rips, L.J. *The psychology of knights and knaves.* Cognition,**31**, 85-116, 1989.

[2] Johnson-Laird, P.N. and Byrne, R.M.J. *Meta-logical puzzles: knights, knaves and Rips.* Cognition, **36**, 69-84, 1990.

[3] Rips, L. *Paralogical reasoning: Evans, Johnson-Laird, and Byrne on liar and truth-teller puzzles,* Cognition, **36**, 291-314, 1990.

[4] Evans, J.St.B.T. *Reasoning with knights and knaves: A discussion of Rips,* Cognition, **36**, 85-90, 1990.

[5] Johnson-Laird, P.N. and Byrne, R.M.J. *Deduction.* Hillsdale: Erlbaum, 1991.

[6] Smullyan, R.M. *What is the Name of This Book? The riddle of Dracula and other logical puzzles.* Englewood Cliffs: NJ:Prentice-Hall, 1978.

[7] Barwise, J. and Etchemendy, J. *The liar: an essay in truth and circularity.* New York: Oxford University Press, 1987.

[8] Dewdney, A.K. *People puzzles: theme and variations.* Scientific American, **260**, **1**, 88-91, 1989.

[9] Rips, L.J. *Cognitive processes in propositional reasoning.* Psychological Review, **90**, 38-71, 1983.

[10] Johnson-Laird, P.N. *Mental Models: Towards a Cognitive Science of Language, Inference and Consciousness.* Cambridge, Cambridge University Press, 1983.

[11] Johnson-Laird, P.N., and Byrne, R. M.J. *Only reasoning.* Journal of Memory and Language, **28**, 313-330, 1989.

[12] Johnson-Laird, P.N., Byrne, R. M.J., and Schaeken, W. *Reasoning by model: the case of propositional inference.* Psychological Review, In press.

[13] Byrne, R.M.J. and Handley, S. *Cognitive processes in meta-deductive strategies.* Submitted manuscript. 1991.

[14] Byrne, R.M.J. *The suppression of valid inferences by conditionals*. Cognition, **31**, 61-83, 1989.

[15] Byrne, R.M.J. and Johnson-Laird, P.N. *Spatial reasoning*. Journal of Memory and Learning, **28**, 564 - 575, 1989.

[16] Byrne, R.M.J. and Johnson-Laird, P.N. *The spontaneous use of propositional connectives*. Quarterly Journal of Experimental Psychology, In press.

[17] Johnson-Laird, P.N., Byrne, R. M.J., and Tabossi, P. *Reasoning by model: the case of multiple qualification*. Psychological Review, **96**, 658 - 673, 1989.

A Conceptual Framework For Semantic Integrity In Large Knowledge-Based Systems

- The World According to GARP ver 1.0 -

Tony Veale, Pádraig Cunningham
Hitachi Dublin Laboratory,
Trinity College
Email: raveale@vax1.tcd.ie

Abstract

Knowledge representation has become recognized universally as the core issue in AI research. This paper describes the fundamental principles of GARP, a conceptual framework for the representation of knowledge in a semantically consistent manner. An argument supporting the necessity of such consistency is presented, with a description of the ways in which this consistency can be achieved and exploited in the field of natural language understanding.

1. Introduction

Current AI research stresses knowledge as the fundamental ingredient in every AI endeavour. Whether it is represented explicitly in a declarative fashion or intricately bound into the workings of a program as procedural heuristics, AI programs cannot function without knowledge. Motivated by this realization, AI researchers are ever more concentrating their efforts on finding the ideal representation of such knowledge, in particular, the qualitative (or naive) world knowledge required to support such diverse areas as NLP, robust expert systems, machine learning and planning. This trend toward very large world knowledge bases was initially sparked by the Naive Physics Manifesto [5] in the late seventies and continues today in the CYC[1] project [6]. Our own work on TWIG[2] [3] centres around the intelligent concept indexing of text and the automatic acquisition of new concepts, a task which is very knowledge intensive and which relies heavily upon the availability of a large semantically rich knowledge base. The larger such knowledge bases become, more urgent becomes the need for the semantic consistency of the knowledge represented within.

---

[1] Note that while CYC is ostensibly a *scruffy* project and the Naive physics Manifesto was originally envisioned as a *neat* endeavour, both have much the same end goal.

[2] TWIG is an intelligent concept acquisition and indexing system, named from the gaelic Tuig - to understand

This paper describes the guiding principles in the design of GARP, the conceptual framework for knowledge consistency employed in the TWIG system. Fundamental to knowledge representation in GARP is the use of constraints in the maintenance of semantic integrity and the recognition of flawed knowledge structures as they are created. Constraints are enforced to trap such illegal structures at entry time, such that the knowledge base is always self consistent.

We first present a brief historical note on semantic integrity, based on McDermott's call for denotation in representation. A justification of our work is then presented via a description of the representational needs of the TWIG system, which also presents an argument for lexical and conceptual integration on the grounds of knowledge consistency. The main thrust of this paper, however, is the implementation of suitable constraints, and their exploitation to the full. It is our thesis that knowledge constraints are themselves rich sources of conceptual information, given a computationally felicitous representation. To this end, we present two diverse applications of knowledge constraints, clearly illustrating the benefits accruing from their use. Overall, we would like this paper to stress the positive aspects of knowledge constraints - not only suited to the role of observant watch-dog but also the natural expression of conceptual relations.

2. Denotational Semantics

Any knowledge representation system which is to be useful for consistently encoding semantics on a reasonable scale must itself have a consistent specification semantics. This is especially so in the case of large knowledge-based systems where the contents of the knowledge base are collated by different agents from possibly different sources. Clearly, the soundness of decisions based upon the semantics of concepts defined in the knowledge base cannot be guaranteed if the concept definitions themselves are not guaranteed to be so. This notion was first introduced by the philosopher Tarski [8] and later introduced to the AI field by Drew McDermott [7] as systematic denotational semantics (SD).

Even a system which employs denotational semantics cannot detect inconsistencies arising from discrepancies between a model and its representation within the system, for its only view of the model is the representation we provide. However, we can at least expect the flawed model to be internally self-consistent. For instance, if we define a Turnip as a member of the class Fruit, the system has no absolute representation (in the platonic sense) of Turnip with which to compare, but it should nevertheless impose the constraint that all instances of Turnip grow above the ground. A model may be misinformed but never unsound. This is the most we can expect from an any integrity maintenance system.

3. Knowledge Base Organization in TWIG

The TWIG knowledge base is organized as a frame-based network of concepts, the backbone of which is a concept IS-A hierarchy. Each distinct concept is represented as a frame, with slot:filler pairs used to store relevant attributes and links to related concept frames. Inheritance of slots is defined across IS-A links only, but other inference techniques such as the use of INVERSE slots are also supported.

The purpose of the concept network in TWIG is twofold. From the user's viewpoint, the network serves as a browsable index of all concepts known to the system (see Figure 1), with appropriate hooks into a hypertext representation of the source document from which the concepts were obtained. Secondly, the network acts as a conceptual knowledge base fundamental to the text understanding and concept acquisition processes. Text understanding is itself a twofold process: first the text is parsed to produce a deep semantic structure which is then used to create a final conceptual representation which is added to the concept network. Conventionally, the main knowledge source employed by the parser is a lexicon (containing dictionary information such as syntax and morphology), while the construction of the conceptual representation clearly utilizes the concept network (a source of encyclopedic knowledge).

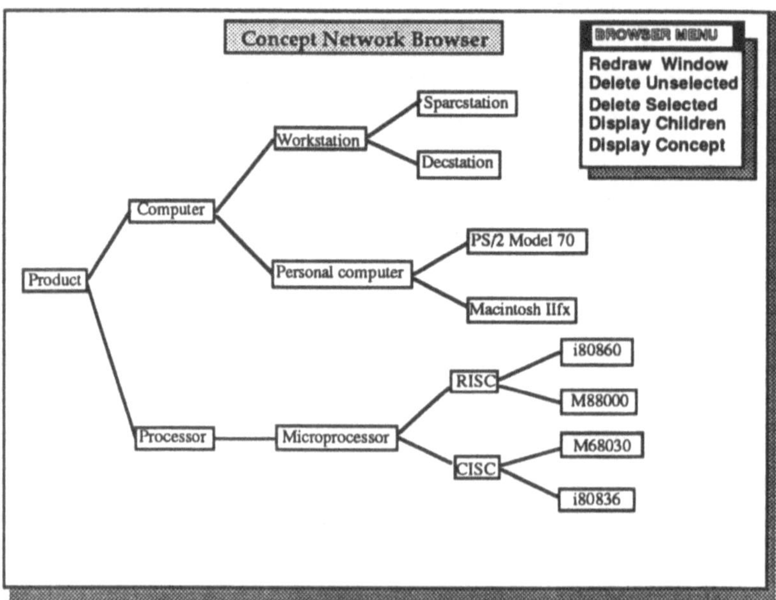

Figure 1 The TWIG Concept Network Browser

An early design decision in the evolution of TWIG prompted the integration of the concept network and the NL lexicon. As an engineering decision, the original motivation was house keeping but the ramifications were to be considerable [4]. The consistency of the knowledge base is endorsed by the consistency of the natural language to which it is tied. It also appears that this approach is a rejection of the modularity argument in linguistics that states that language ability is inate and can be separated from other cognitive processes.

3.1 Lexical & Conceptual Integration

One of the fundamental tenets in the design of the TWIG system is that our two major knowledge sources, the lexicon and concept network, should be one and the same. With such an integration, each concept corresponds to a word in the host language , with syntactic, semantic and conceptual information each represented at the same level of abstraction.

Intuitively, there is a consistency to be gained from such an integration. An integrated knowledge base may use slot names and filler values which are themselves concepts with associated frames; this is nothing new - it is encouraged for the sake of consistency in most KR systems (such as CYC). However, each slot name and filler value will also have a corresponding lexical definition with associated *how to* knowledge in the text processing domain. We would therefore suggest that such concepts have a deeper representation than their bipartite equivalents. In a system that supports a distinct lexicon and concept network, it is possible for conceptual information to be *lost* in the network, for if the lexicon provides no explicit mapping onto a particular concept, the system has no model within which to provide an interpretation. For instance, imagine a concept Liquid-Nitrogen which has the slot:filler pair Temperature-of = Freezing. The temptation is of course to assume the system understands that the temperature of liquid nitrogen is at freezing point, but such an assumption is not at all supported by the definition. We first have to show that the slot name temperature-of actually represents the temperature concept, which may be apparent to us, but certainly not to the system. Without an explicit relation, the interpretation has to be provided by the clients of the knowledge base, relegating the system's understanding of its own contents to shallow and opaque. In an integrated system, however, each concept is intrinsically bound to its usage in the host language, which serves as a model against which the knowledge base can interpret its own contents. Again appealing to intuition, the larger such a knowledge base grows, the more consistent it becomes, as the possible vocabulary of its representation is fleshed out. In effect, the concept network gradually assumes the consistency of the host language.

4 Knowledge Specification and Constraints

In a frame-based KR system as outlined above, the nature of knowledge constraints is much simplified. Under such a scheme, knowledge is represented as a collection of frame structures, which are themselves composed of slots and associated fillers. Therefore, to fully constrain a frame based system, it simply suffices to constrain the slot names associated with a frame and the filler values associated with those slots.

In GARP, this first constraint manifests itself as slot declaration - a slot must be declared for a particular frame if it is to be bound with a filler within that frame. Slot declarations are inherited from superclass definitions, which effectively act as templates for the generation of subclasses and instances. Note that this does not necessarily constrain the information expressed in a class definition, rather the nature in which it is expressed. Slot declaration prohibits the introduction of new slot:filler pairs *on the fly* , much like the Pascal programming language (as opposed to BASIC) has disdain for variable names which have not been declared in advance. The second constraint is satisfied by introducing a check for semantic compatibility between slot names and slot fillers. For instance, one should not be permitted to fill a Colour slot with the value Big, or indeed a Size slot with the value Red.

4.1 Implementing Knowledge Constraints

Having outlined the nature of the necessary constraint types for knowledge base consistency, we shall now turn our attention to the implementation of such constraints in a frame-based system such as GARP. Essentially, constraints upon a frame-based system restrict membership of the set of permissible <frame:slot:filler> triples to those triples which are deemed semantically congruous relevant to the model under representation. Thus, a most suitable form for such constraints to assume is that of a <frame:slot:filler> triple (denoted F.S.x) - ie, slot constraints should themselves be implemented as slots.

Slot declarations most naturally assume the form of *template* slots with *wildcard* fillers; GARP employs the UNIX wildcard convention wherein ? indicates a single value and * indicates one or more possible values. Naturally, no constraints are placed on slot declarations themselves, and the user can define as many wildcard bearing slots as he so wishes. When defining a slot which contains a concrete (non wildcard) filler list, the system first determines whether the frame in question inherits from its superclasses a slot of the same name containing the appropriate wildcard. Clearly, slots containing more than one filler must inherit the * wildcard, while the ? permits at most one legal filler. In the case of the * wildcard, a * is automatically added to the list of local fillers to allow for future expansion in the host slot.

1. Slot Declaration

F.S.x is legally declared if

exists F.S.*

OR

exists F.S.? and $|x| = 1$

In addition to the ? and * wildcards, GARP also supports the template slot filler SELF. Whenever a new frame is defined, the KB manager automatically includes a slot of the same name containing the SELF filler. This allows the clients of the knowledge base to differentiate between those concepts that subsume a particular entity and those concepts which are merely supported by that entity. For instance, the entity Bill supports the concepts Colour, Size and Age while being subsumed by the concepts Man, Human, Mammal etc. Essentially, SELF slots provide a simple test for concept subsumption, or effective class membership.

2. Concept Support:

concept X supports concept Y if exists X.Y.? or exists X.Y.*

3. Concept Subsumption:

concept X subsumes concept Y if exists Y.X.SELF

In turn, this test for concept subsumption enables us to enforce the second constraint type, that of slot name and slot filler compatibility. Essentially, every slot name and slot filler must have an associated concept (frame) definition, and every slot concept must subsume its associated filler concepts. (again, we exclude the cases where the slot is declarative and contains a wildcard, or indeed where it contains non-conceptual data such as lists, strings or numbers).

4. Slot : Filler Compatibility

A slot S and filler F are compatible if exists F.S.SELF

We can summarize both constraint types as follows: No frame can support a slot which has not been declared as applicable by a superclass, and no slot can contain a filler which it does not subsume.

5. Exploiting Knowledge Constraints

Primarily, the constraint type most open to exploitation is the enforced declaration of slots. Such slot declarations can be employed to determine the conceptual link between two concepts. In effect, slot declarations collectively act as templates, which guide the composition of related concepts for the automatic construction of complex knowledge structures (as needed in NLP and ML). The automatic reinforcement of slot:filler compatibility can likewise serve as the basis of a preference semantics system [9], where the name of a slot may also be used as a default filler value in the absence of specified fillers. This follows from the constraint that the concept associated with a slot name must subsume the contents of that slot.

5.1 Determination of Conceptual Relations Via Constraints

We can classify conceptual relations into two distinct types: Subsumption relations and Support relations. A Subsumption relation exists between two concepts when one concept subsumes the other, such as Vehicle : Car and Apple : Fruit. A Support relation exists between two concepts when one concept can legally fill a slot of the other, (obeying the outlined KB constraints) as in Apple : Red and Macintosh IIfx : M68020.

A Support relation exists between concepts X and Y if Supports (X,Y) OR Supports (Y,X). We define Supports(X,Y) as follows : A concept X supports a concept Y if a legal slot S of X subsumes Y (ie a slot of X can legally contain Y). Thus, determination of Support (X,Y) is equivalent to finding a slot S of X (possibly inherited) which is a superclass of Y. The concept S is thus the conceptual relation linking the concepts X and Y. For instance, Supports(Apple, Red) = Colour and Supports(Macintosh IIfx, M68030) = Processor. The natural choice of implementation algorithm is that of marker passing [1], otherwise known as Spreading Activation [2]. Marker passing would proceed recursively from both patriarch concepts, traversing the IS-A backbone of the concept network (along bold lines in Figure 2) tagging each node encountered with a token, @SUPER say. For every concept node visited, the concepts associated with that node's constituent slot names are also marked with a @SUPPORT marker (see dashed lines in Figure 2) Any concept marked with both @SUPER and @SUPPORT tokens represents the conceptual relation that exists between both patriarch concepts.

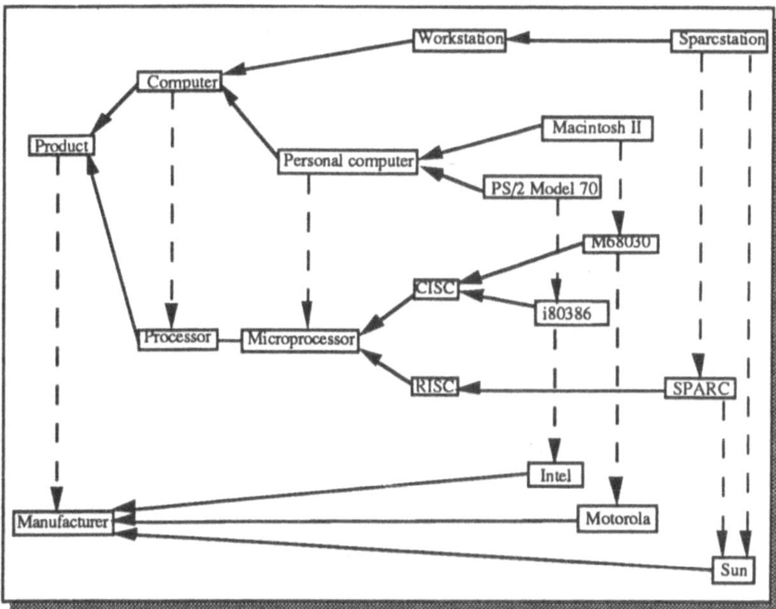

Figure 2 - Concept Network with Subsumption and Support Links illustrated

Unfortunately, marker passing is not a cost-effective (timewise) search strategy for the above task, especially in the case of very large knowledge bases. Such an overhead may be tolerable when additional evidence supports a possible conceptual relation between two concepts, but we suspect that in the vast majority of cases, such evidence would be lacking. No, what is clearly needed is an additional constraint which effectively prunes the search space of the marker passing process. One such constraint which we are currently evaluating within the TWIG knowledge base restricts the descriptive scope of each concept to at most one relation. For instance, the Red concept always prescribes the Colour relation, no matter what concept it is used to describe; the M68030 and i386 concepts always fill the processor slot of the host concept, and so on. In effect, this constraint eradicates any variance among inter-concept relations in the GARP knowledge base.

We can now significantly improve the procedure for determining Supports(X,Y). Each GARP concept is assumed to contain a Describes slot, which prescribes the type of conceptual relation in which the concept may partake (ie it contains the name of the slot in which the concept normally resides). For instance, the Describes slot of Colour contains the filler value Colour, which is inherited by the concepts Red, Green, Blue and so on. Thus, Red always partakes in the Colour relation, and resides in the Colour slot of the concept it is used to describe. In turn, the Company concept prescribes the Manufacturer relation, a slot which is supported by subclasses of the Product concept.

Thus, Supports(M68030, Motorola) = Manufacturer. The following flowchart (figure 3) illustrates the process of support relation determination.

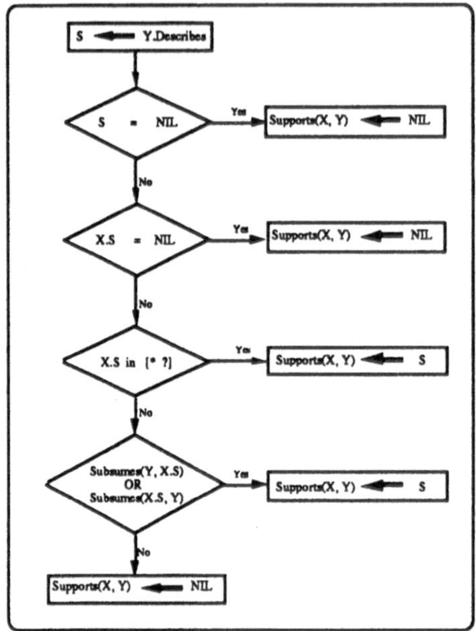

Figure 3 - Determination of Supports(X, Y)

The algorithm outlined in figure 3 operates as follows: the conceptual relation linking two concepts X and Y (if any such relation exists) is first determined by evaluating Describes(Y), simply by accessing the Describes slot of Y. Given this relation (stored in S), we determine whether X has a corresponding slot S to accommodate Y. If so, we need to determine whether the filler of this slot (either a wildcard or a concept) is semantically compatible with Y. For the purposes of this algorithm, two distinct concepts are deemed compatible if a Subsumption relation exists between them. For example, Company and Apple-inc are compatible, IBM and Company are compatible, but Apple-inc and IBM are not. Thus, Supports(IBM, Proprinter) = Printer but Supports(IBM, Laserwriter) = NIL, as Laserwriter.manufacturer = Apple.

Note also that in this implementation, slot lookup has replaced spreading activation as the main search strategy. In the full blown marker passing scheme, activation is spread from both patriarch concepts, and whenever both *wave fronts* collide the system performs a check for a valid conceptual relation. With this algorithm, however, the first activation wave is accomplished by evaluating Y.Describes, and the second by evaluating X.S. In effect, we prune the search space by providing direction to the search process. The cost, however, is a restriction on relational variance, the ramifications of which have yet to be fully evaluated in the TWIG system.

6 Knowledge Constraints in Action

Having outlined the form, implementation and possible use of knowledge constraints, we now focus our attention on some example applications which exploit these constraints to their fullest. We present here a brief overview of two such practical uses in the areas of NLP and knowledge specification.

6.1 Natural Language Processing And Semantic Parsing

As stated earlier, the main purpose of the TWIG system is knowledge extraction from text, a process which clearly requires the power of a natural language processor. Without describing the present NLP module in full, we present here an outline of the pragmatics component, which uses an agenda mechanism to perform anaphoric/referent resolution and the composition of related concepts into representative meaning structures. The workings of this particular agenda mechanism rely heavily on the ability to determine the conceptual relation linking two distinct concepts, which in turn relies on the fullest exploitation of knowledge constraints (as described in section 5).

The pragmatics agenda mechanism employs a reference-ordered stack (the agenda) to track text referents during the understanding process. Each time the NL analyser encounters a new referent, the agenda is searched (most recent additions first) for a matching referent which has previously been encountered. The matching process is performed on the basis of subsumption relations and attribute compatibility (an Apple should match a Fruit, but a Green Apple should not match a Red fruit). If a suitable match is found, it is removed from the agenda, unified with the trial referent, and pushed onto the top of the agenda. If a match is not found, the trial referent is simply pushed onto the top of the agenda. In this way, the agenda always stores referents in the order in which they are most likely to be referenced. Clearly, referent matching is very much dependent on the ability to determine subsumption relations among concepts, as provided by the slot declaration constraint.

In addition, the agenda mechanism can be augmented to perform concept composition, by again exploiting slot declaration constraints in the determination of conceptual relations. While searching the agenda for a matching referent, TWIG does not merely perform a check for concept subsumption, rather a more general test for concept connectivity, which involves a check for both subsumption and support. In this way, referents residing on the agenda can be linked together via the appropriate conceptual relations, enabling the creation of larger, more complex (and thus, more representative) meaning structures.

Indeed, such a mechanism usually forms the core of semantic parsing systems, which exhibit little regard for syntax and prefer instead to rely on conceptual information for linking related concepts. Augmented with a noun phrase preprocessor, the above agenda mechanism could well be used for *skimming* based language processors. In fact, while the TWIG NL component employs a chart parser in conjunction with a unification grammar, a skimming mode is provided for *speed reading* of texts. Running in this mode, the chart parser is disconnected and the pragmatics component receives its input directly from the NP preprocessor.

Figure 4 provides a snapshot of the agenda mechanism in action, where full chart parsing has been enabled. The diagram demonstrates TWIG operating on a product review fragment taken from BYTE magazine. Illustrated are the processes of concept composition and referent resolution, which act together to link related concepts into larger representative structures for indexing. Illustrated also is the *Filing Cabinet* metaphor of concept indexing - those concepts marked with a cabinet icon are actually indexed for future retrieval, while others are filtered and simply *washed out* of the system.

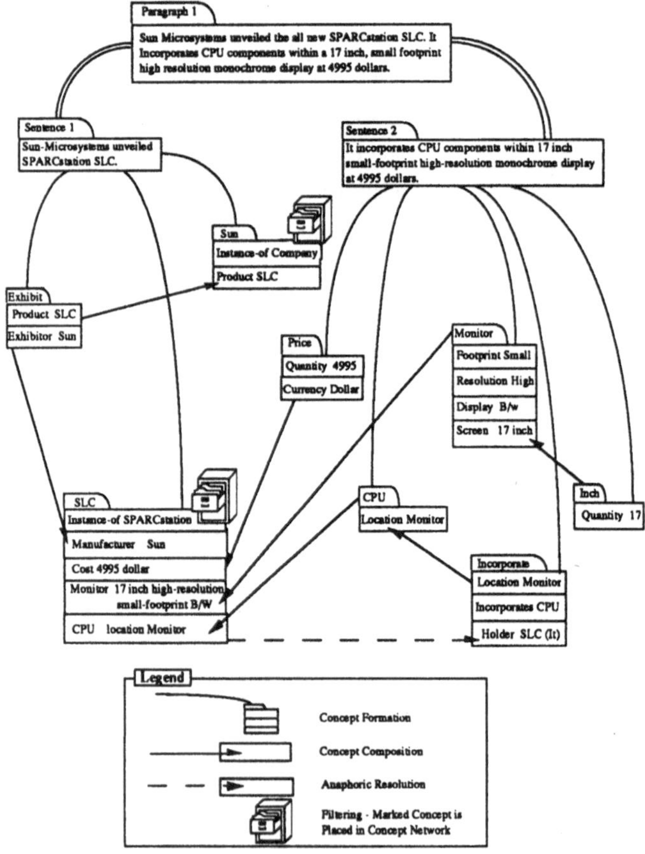

Figure 4 - NL understanding in TWIG

The diagram illustrates the complete understanding process from input text to output concept structures. The particular text used is taken from a BYTE product review, TWIG's current experimental domain of interest.

6.2 An Intelligent Knowledge-Base Editor

Although this application is somewhat lightweight in comparison to the NL domain presented above, it does illustrate nicely the role of knowledge constraints (in particular, their use in determination of conceptual relations) in simulating intelligent behaviour. The proposed scenario is as follows: having constructed a large knowledge base extensive enough to allow TWIG to operate competently on unrestricted text, the system is released for commercial use. An interface is provided to allow the end-user to augment the knowledge base / concept network interactively, such that said user may customize TWIG to the requirements of his particular domain.

Rather than force the user to create new concepts via frame definitions, which only serve to complicate the entry process, a graphical interface is supplied. The user simply enters the name of the new concept in the concept network browser window, and proceeds to draw links to related concepts in the network. The nature of these links is not explicitly specified by the user, rather link types are inferred by determining the conceptual relation connecting the corresponding concepts. Naturally, the first link entered by the user is deemed to represent a Subsumption relation, as the new entry has yet to be given any conceptual information.

This process is illustrated in Figure 5, which demonstrates the addition of a new concept Macintosh-IIfx to the concept network (and thus, the lexicon). Having been prompted for the name of the new entry, the user proceeds to draw three links (shown as dotted lines) to the existing concepts which best define the entry: Macintosh-II, Motorola and M68030. TWIG naturally assumes that the first link specified always indicates a subsumption relation, and automatically defines Macintosh-IIfx as a subclass of the concept Macintosh-II. Employing the algorithm outlined in section 5, TWIG proceeds to determine Supports(Macintosh-IIfx, Motorola) = Manufacturer and Supports(Macintosh-IIfx,M68030) = Processor. It therefore completes the automatic concept definition by asserting the slots <Macintosh-IIfx : Manufacturer> = Motorola and <Macintosh-IIfx : Processor> = M68030.

Intelligent linking shields the user from the internal vagaries of concept definitions - in effect, they act as black boxes. The user need never know specifically how some particular piece of information is represented (such as specific slot names); it is sufficient to know how the concept interacts with others. We believes this adheres closely to the object oriented paradigm of knowledge representation.

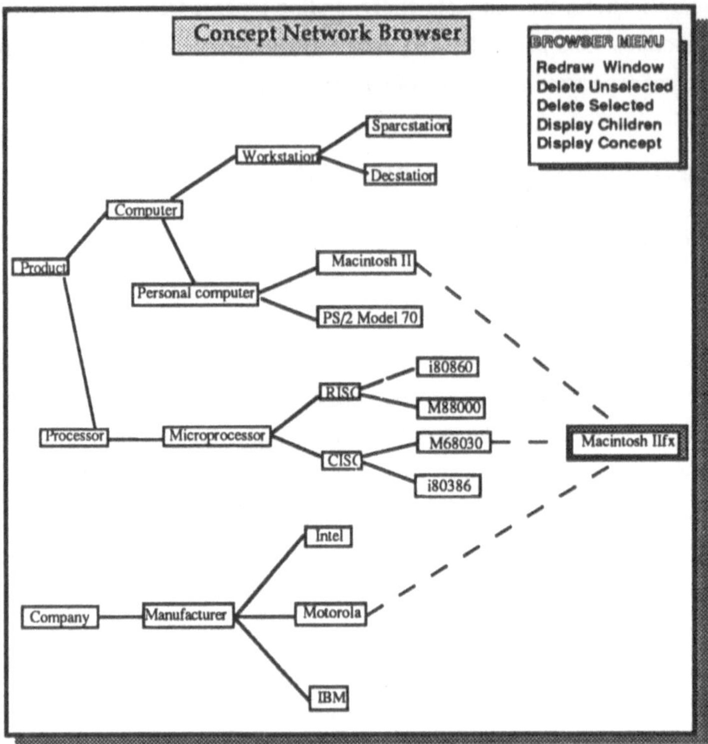

Figure 5 The Intelligent Concept Network Editor.
Illustrated is the Creation of the new concept Macintosh IIfx

7 Constraints As Knowledge Advisers

Thus far this paper has concentrated on the presentation of knowledge constraints as both necessary consistency measures and a useful source of knowledge in themselves. Constraints restrict the set of possible <frame:slot:filler> triples to that collection which is semantically congruous, relative to the model under interpretation: each slot definition is judged by these constraints and deemed either legal or illegal, black or white. However, in many cases, constraints can also serve in the somewhat grey role of watchful adviser, indicating to the concept designer just why a particular definition is illegal, and what the designer can do to make amends.

As we have stressed throughout this paper, GARP ensures that every slot filler is subsumed by the corresponding slot name - ie, an IS-A relation exists between every slot concept and corresponding filler concept. We believe this to be a valid representation constraint. However, many users may actually find this constraint to restrictive. Consider the example CYC definition given in [6], where the concept DanishFurniture is given the slot <DanishFurniture : Manufacturer> = Danish. Clearly, GARP would reject such a definition, as the concept Danish has not been defined as a Manufacturer in the concept network. However, it is also clear (to us) that the user intended to define the nationality of the Manufacturer as Danish. The correct approach would thus define <DanishFurniture : Manufacturer> = DanishManufacturer and <DanishManufacturer : Nationality> = Danish (see Figure 6). This illustrates a major problem of knowledge specification - it is all to easy to blur the distinction between Support and Subsumption relations when linking concepts together. By the same token, however, it should be a matter of equal ease for GARP to detect such knowledge *sublimation,* as a Support relation exists between the slot and its illegal filler. Such information is sufficient to enable GARP to offer the above advice, and perhaps correct the error automatically (with the user's prior consent, of course).

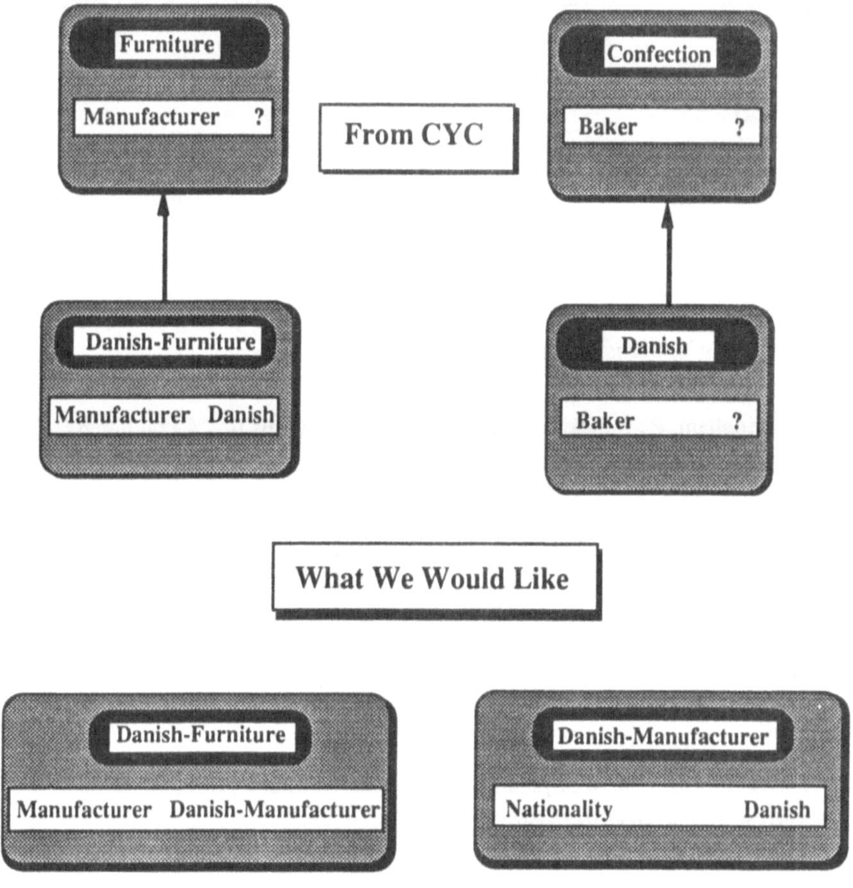

Figure 6: How to make Danish Furniture

8 Summary and Conclusions

AI literature of recent years has stressed the importance of knowledge, and its appropriate representation, as the foundation of any AI endeavour. In accordance with this view, this paper additionally stresses the need for the semantic consistency of such representation, while providing arguments for the utility of such constraints as contributory knowledge sources in themselves. We have presented a homogeneous expression of semantic constraints, where the representation of constraints mirrors the form of the knowledge itself. All conceptual constraints can be reduced to restrictions upon slot form, so it is appropriate that such constraints should themselves assume slot form. Constraints and the knowledge they shape are cut from the same cloth. We argue that this homology can be exploited as a source of intelligent behaviour and describe several ready applications of this *latent* knowledge. The use of constraints is the stick that governs the knowledge base; the exploitation of these constraints is the carrot.

References

[1] Charniak Eugene, "A Neat Theory of Marker Passing", *Proc. of the National Conference on Artificial Intelligence* AAAI '86, 1986

[2] Collins A., Loftus E.F., "A Spreading Activation Theory of Semantic Processing", *Psychological Review,* 1975

[3] Cunningham, P., Fujisawa H., Hederman L., Cummins F., "A combined approach to text retrieval using concept networks linked to hypertext", in *Proceedings of the First Japanese Knowledge Acquisition for Knowledge-Based Systems Workshop* 1990

[4] Cunningham, P., Veale, T. "Organizational Issues arising from the Integration of the Concept Network and Lexicon in a Text Understanding System", presented at the International Joint Conference on Artificial Intelligence 1991

[5] Hayes, P., "The Naive Physics Manifesto" in (D. Michie, ed.) *Expert Systems in the microelectronic age,* Edinburgh Univ. Press, 1978

[6] Lenat, D., Guha, R.V., *Building Large Knowledge-Based Systems,* Addison Wesley 1990 ISBN 0-201-51752-3

[7] McDermott, D., "Tarskian Semantics, or No Notation Without Denotation!", *Cognitive Science* **2**(3), 1978

[8] Tarski, A. "Der Wahrheitsbegriff in den formalisierten Sprachen", *studia Philos*, 1936

[9] Wilks, Y., "A Preferential, Pattern-Seeking, Semantics for Natural Language Inference", *Artificial Intelligence* 6, 1975

Generalized Knowledge Representation using Free Logic

James Bowen
Dept. of Computer Science, North Carolina State University
Raleigh, NC 27695-8206, USA

Dennis Bahler
Dept. of Computer Science, North Carolina State University
lsize Raleigh, NC 27695-8206, USA

Abstract

Declarative rule-based systems can be generalized to constraint-based systems. However, although conventional constraint systems require that the set of parameters which exist in a problem be known *ab initio*, there are some applications in which the existence of certain parameters is dependent on conditions whose truth or falsity can only be determined dynamically. In this paper, we show how this conditional existence of parameters can be handled in a mathematically well-founded fashion by viewing a constraint network as a set of sentences in free logic. Based on these ideas, we have developed, implemented and applied to a range of applications, a constraint language in which any sentence in full first-order free logic, about a many-sorted universe of discourse which subsumes the real numbers, is a well-formed constraint.

1 Introduction

A program in a declarative rule-based system is a theory written in a first-order language $\mathcal{L} = \langle \mathcal{P}, \mathcal{F}, \mathcal{K} \rangle$.[1] Rule-based programming systems restrict in various ways the well-formed sentences (or rules) that may appear in a theory (or program). By definition, rule-based systems admit only single-consequent implication statements. Thus, for example, they cannot handle disjunctive knowledge [13]. Most such systems also restrict the allowable sentences to the ground sentences, while those that do allow quantification usually only admit implicit quantification – in rules that contain logic variables, the variables are understood to be universally quantified, in prenex normal form.

As will be shown below, a conventional constraint network corresponds to a theory in first-order classical logic (FOCL) and constraint networks corresponding to arbitrary FOCL theories can be specified. Thus, constraint-based programming systems can be more expressive than rule-based programming systems. However, although conventional constraint systems [10] require that the set of parameters which exist in a problem be known *ab initio*, there are some applications [6, 11] in which not just the value of certain parameters, but also their very existence, depends on conditions whose truth can only be determined dynamically. In this paper, we show how this conditional existence of parameters can be handled in a mathematically well-founded fashion by viewing a constraint network as a set of sentences in first-order free logic (FOFL) [9]. As well as having a better-developed theoretical

[1] In the language $\mathcal{L} = \langle \mathcal{P}, \mathcal{F}, \mathcal{K} \rangle$, $\mathcal{P}$ is a set of relation or predicate symbols, $\mathcal{F}$ is a set of function symbols and $\mathcal{K}$ is a set of object or constant symbols.

underpinning, this approach is more general than that presented in [11]. We briefly review a language, based on these ideas, which we have developed, implemented and applied to a range of applications.

2 Conventional Constraint Processing and Classical Logic

A constraint is a declarative specification of some restriction that applies to the values which may assumed by a collection of one or more parameters. A set of constraints form a network by virtue of sharing parameters. Formally, a constraint network may be defined as follows:

Constraint Network:
A constraint network is a triple $\langle \mathcal{U}, \mathbf{X}, \mathbf{C} \rangle$, where $\mathcal{U}$ is a universe of discourse, $\mathbf{X}$ is a finite tuple of q non-recurring parameters, and $\mathbf{C}$ is a finite set of r constraints. In each constraint $C_k(T_k) \in \mathbf{C}$, T_k is a sub-tuple of $\mathbf{X}$, of arity a_k, and $C_k(T_k)$ is a subset of the a_k-ary Cartesian product $\mathcal{U}^{a_k}$.

Thus, the parameters in $\mathbf{X}$ take their values from the universe of discourse $\mathcal{U}$ subject to the constraints in $\mathbf{C}$. The overall network constitutes an intensional specification of a joint possibility distribution for the values of the parameters in $\mathbf{X}$. This distribution is a q-ary relation on $\mathcal{U}^q$, called the *intent* of the network:

The Intent of a Constraint Network:
The intent of a constraint network $\langle \mathcal{U}, \mathbf{X}, \mathbf{C} \rangle$ is
$$\Pi_{\mathcal{U},\mathbf{X},\mathbf{C}} = \overline{C}_1(\mathbf{X}) \cap \ldots \cap \overline{C}_r(\mathbf{X}),$$
where, for each constraint $C_k(T_k) \in \mathbf{C}$, $\overline{C}_k(\mathbf{X})$ is its cylindrical extension [7] in $\mathcal{U}^q$, the Cartesian product which forms the value space for $\mathbf{X}$.

Based on this, several forms of constraint satisfaction problem (CSP) can be defined [3]. However, we will concern ourselves here with just the Exemplification CSP, which can be defined as follows:

The Exemplification CSP:
Given a network $\langle \mathcal{U}, \mathbf{X}, \mathbf{C} \rangle$, return some tuple from $\Pi_{\mathcal{U},\mathbf{X},\mathbf{C}}$, if $\Pi_{\mathcal{U},\mathbf{X},\mathbf{C}}$ is non-empty; otherwise, return nil.

Thus, solving the Exemplification CSP involves determining whether there can be any valuation for all the parameters in the network which satisfies all the constraints. If there can be such a valuation, solving the Exemplification CSP involves finding one such valuation.

Solving the Exemplification CSP for conventional constraint networks may be regarded as semantic modeling in first-order classical logic, in the following sense. The constraints in $\mathbf{C}$ correspond to sentences of an FOCL theory Γ, written in some first-order language $\mathcal{L} = \langle \mathcal{P}, \mathcal{F}, \mathcal{K} \rangle$. The parameters in $\mathbf{X}$ correspond to constant symbols, from $\mathcal{K}$, which appear in Γ. Solving the Exemplification CSP corresponds to finding an interpretation function $\mathcal{I}$ from the symbols of $\mathcal{L}$ to entities in, and relations over, $\mathcal{U}$, such that Γ is satisfied under the model $\mathcal{M} = \langle \mathcal{U}, \mathcal{I} le$ of the language $\mathcal{L}$.

2.1 Example 1

Consider the task of semantically modeling the theory $\Gamma = \{positive(area), nonnegative(velocity), nonnegative(volume), volume = area * velocity, velocity \leq 200\}$, written in the first-order language $\mathcal{L} = \langle\{positive, nonnegative, =, \leq\}, \{*\}, \mathcal{R} \cup \{area, velocity, volume\}\rangle$.

In this language, $\mathcal{R}$ is the set of constant symbols composed from the characters +, −, . and 0..9 according to a grammar for real numeric strings. $\mathcal{R}$ is distinguished from $\Re$, the set of real numbers. In this paper, to distinguish between symbols of $\mathcal{L}$ and entities of $\mathcal{U}$, we use typewriter font for the latter. Thus, $200 \in \mathcal{R}$ is a constant symbol while $\texttt{200} \in \Re$ would be in a universe of discourse.

Suppose we are given a universe of discourse $\mathcal{U} = \Re$ and a partial interpretation function $\mathcal{I}_p$ for $\mathcal{L}$, which provides an interpretation for each predicate and function symbol of $\mathcal{L}$ as well as a bijective mapping between $\mathcal{R}$ and the finitely expressible rationals Q_f, where $Q_f \subset Q \subset \Re$. That is, as well as constant mappings of the form $200 \mapsto \texttt{200}$, $\mathcal{I}_p$ contains these predicate and function mappings:

$$\begin{array}{lcl} positive & \mapsto & \Re^{+} \\ nonnegative & \mapsto & \Re^{0+} \\ = & \mapsto & \{\langle X,Y\rangle | X \in \mathcal{U} \wedge Y \in \mathcal{U}\ edge\ \textsc{equals}(X,Y)\} \\ \leq & \mapsto & \{\langle X,Y\rangle | X \in \Re \wedge Y \in \Re \wedge \textsc{leq}(X,Y)\} \\ * & \mapsto & \{\langle X,Y,Z\rangle | X \in \Re \wedge Y \in \Re \wedge \\ & & \qquad Z \in \Re \wedge\ \textsc{equals}(Z,\textsc{times}(X,Y))\}. \end{array}$$

The constraint network $\langle \mathcal{U}, \mathbf{X}, \mathbf{C}\rangle$ corresponding to this situation is a possibility distribution for interpretations of the remaining uninterpreted constant symbols of $\mathcal{L}$, such that Γ is satisfied. The components of this network are: $\mathcal{U} = \Re$, $\mathbf{X} = \langle area, velocity, volume\rangle$ and $\mathbf{C} = \{C_1(area), C_2(velocity), C_3(volume), C_4(area, velocity, volume), C_5(velocity)\}$. The constraints are defined as follows:

$$\begin{array}{l} C_1(area) = \Re^{+} \\ C_2(velocity) = \Re^{0+} \\ C_3(volume) = \Re^{0+} \\ C_4(area, velocity, volume) = \\ \qquad \{\langle X,Y,Z\rangle | X \in \Re \wedge Y \in \Re \wedge Z \in \mathcal{U} \wedge \textsc{equals}(Z,\textsc{times}(X,Y))\} \\ C_5(velocity) = \{X | X \in \Re \wedge \textsc{leq}(X,\texttt{200})\}. \end{array}$$

Each sentence in the theory has a corresponding constraint in the network, the definition of which depends on whatever information is provided, by the partial interpretation $\mathcal{I}_p$, about the symbols appearing in the sentence. Consider, for example, $C_4(area, velocity, volume)$, which corresponds to the sentence $volume = area * velocity$. There are three parameters in this constraint, corresponding to the three constant symbols in the sentence. The restriction imposed by the constraint is derived from the interpretations in $\mathcal{I}_p$ for the symbols = and * in the sentence. Consider $C_5(velocity)$, which corresponds to the sentence $velocity \leq 200$. Although two constant symbols appear in the sentence, there is only one parameter in the constraint, because $\mathcal{I}_p$ provides an interpretation for 200.

The cylindrical extensions of these constraints, into the Cartesian product which represents the value space for the tuple $\langle area, velocity, volume\rangle$, are as follows:

$\overline{C}_1(area, velocity, volume) = \Re^{+} \times \mathcal{U} \times \mathcal{U}$
$\overline{C}_2(area, velocity, volume) = \mathcal{U} \times \Re^{0+} \times \mathcal{U}$
$\overline{C}_3(area, velocity, volume) = \mathcal{U} \times \mathcal{U} \times \Re^{0+}$
$\overline{C}_4(area, velocity, volume) =$
$\{\langle X, Y, Z\rangle | X \in \Re \wedge Y \in \Re \wedge Z \in \mathcal{U} \wedge \text{EQUALS}(Z, \text{TIMES}(X, Y))\}$
$\overline{C}_5(area, velocity, volume) = \mathcal{U} \times \{X | X \in \Re \wedge \text{LEQ}(X, 200)\} \times \mathcal{U}.$

Intersecting these cylindrical extensions, we get the following as the intent of the network:

$$\Pi_{\mathcal{U},\mathbf{X},\mathbf{C}} = \{\langle X, Y, Z\rangle | X \in \Re^{+} \wedge Y \in \Re^{0+} \wedge \text{LEQ}(Y, 200) \wedge Z \in \Re^{0+} \wedge \text{EQUALS}(Z, \text{TIMES}(X, Y))\}.$$

This intent is not the empty set and any constituent tuple, such as ⟨100, 150, 15000⟩, is a solution to the Exemplification CSP based on the network.

2.2 Example 2

Constraint networks and constraint processing have wide applicability, in areas as diverse as design [4] and computer vision [12]. The network in Example 1 represents some considerations affecting the cross-sectional *area* of a pipe, the *volume* per second of fluid passing through the pipe and the *velocity* of the fluid.

The theory in Example 1 contained only ground atomic sentences. However, constraint networks corresponding to the task of completing the semantic model for arbitrary FOCL theories can be specified.[2] An example FOCL theory complex enough to be truly interesting is beyond the scope of this paper (but see [6]). However, for the sake of example, consider the following.

Suppose that there is a set of standard areas and that it is required that the pipe in Example 1 must have a cross-sectional area smaller than at least two of these standard areas. This additional requirement could be specified by adding the following sentence to the theory in Example 1:

$$(\exists X)(\exists Y)(standard(X) \wedge standard(Y) \wedge \neg(X = Y) \wedge area < X \wedge area < Y).$$

Thus, in this second example, the following theory would result:

$$\begin{aligned}\Gamma = \{&positive(area),\ nonnegative(velocity),\ nonnegative(volume),\\ &volume = area * velocity,\ velocity \leq 200,\\ &(\exists X)(\exists Y)(standard(X) \wedge standard(Y) \wedge \neg(X = Y) \wedge area < X \wedge area < Y)\}.\end{aligned}$$

In addition, the language $\mathcal{L}$ in Example 1 would be extended, in this second example, to include the predicate symbols *standard* and <, while the partial interpretation function $\mathcal{I}_p$ would be extended to include a mapping from the predicate symbol *standard* to the set of standard areas and from the predicate symbol < to the usual sense of this symbol when applied to the reals.

Suppose, for example, that these predicates were interpreted as

$$\begin{aligned} standard &\mapsto \{46.9, 57.8, 95.7, 124.7\}\\ < &\mapsto \{\langle X, Y\rangle | X \in \Re \wedge Y \in \Re \wedge \text{LESS}(X, Y)\}.\end{aligned}$$

[2] Of course, because of Godel's theorem on the undecidability of FOCL, it cannot be guaranteed that the satisfiability or otherwise of arbitrary networks can be determined by any realisable algorithm.

Then, the constraint corresponding to the sentence $(\exists X)(\exists Y)(standard(X) \wedge standard(Y) \wedge \neg(X = Y) \wedge area < X \wedge area < Y)$ would be

$$C_6(area) = \{X | X \in \Re \wedge \text{LESS}(X, 95.7)\}.$$

The intent of the resultant network would be

$$\Pi_{\mathcal{U},\mathbf{X},\mathbf{C}} = \{\langle X, Y, Z\rangle | X \in \Re^{+} \wedge \text{LESS}(X, 95.7)\} \wedge Y \in \Re^{0+} \wedge \text{LEQ}(Y, 200) \wedge Z \in \Re^{0+} \wedge \text{EQUALS}(Z, \text{TIMES}(X, Y))\}.$$

2.3 Example 3: Inadequacy of Conventional Networks

Although conventional constraint networks are adequate in many situations, there are also many practical problems for which the expressive competence provided by FOCL is not adequate to facilitate natural problem specifications. Consider, for example, the following.

Extend Example 2 to incorporate considerations about the cross-section of the pipe. Circular and rectangular shapes are allowed; circular pipes have a cross-sectional radius; rectangular pipes have a breadth and height.

This example is beyond the expressive competence of conventional constraint networks, because the *radius* cannot exist if the *height* and *breadth* exist. We have encountered a fundamental ontological inadequacy of classical logic: its inadequacy for reasoning about existence or non-existence of the objects denoted by constant symbols [8]. In what follows, we show how this difficulty can be overcome, by using FOFL [9], rather than FOCL, as the theoretical basis for constraint networks.

3 Free Logic

In classical logic, a model $\mathcal{M} = \langle \mathcal{U}, \mathcal{I} \rangle$ for a first-order language $\mathcal{L} = \langle \mathcal{P}, \mathcal{F}, \mathcal{K} \rangle$ comprises a universe $\mathcal{U}$ and an interpretation function $\mathcal{I}$. The interpretation function $\mathcal{I}$ assigns to each predicate symbol $p \in \mathcal{P}$ and each function symbol $f \in \mathcal{F}$ some relation $\mathcal{I}(p)$ or $\mathcal{I}(f)$ over $\mathcal{U}$. In addition, the interpretation function assigns to each constant symbol $\kappa \in \mathcal{K}$ some element $u \in \mathcal{U}$.

Free logic [9] differs from classical logic in that a first-order free logic language $\mathcal{L} = \langle \Omega, \mathcal{P}, \mathcal{F}, \mathcal{K} \rangle$ contains a distinguished unary predicate symbol $\Omega \notin \mathcal{P}$ and a free logic model $\mathcal{M} = \langle \mathcal{U}, \mathcal{I}, \mathcal{S} \rangle$ contains a third item, a story $\mathcal{S}$. As before, the function $\mathcal{I}$ assigns to each $p \in \mathcal{P}$ and each $f \in \mathcal{F}$ a relation over $\mathcal{U}$. A key difference, however, is that the mapping from $\mathcal{K}$ to $\mathcal{U}$ need not be total: the interpretation function $\mathcal{I}_p$ need not assign a $u \in \mathcal{U}$ to every $\kappa \in \mathcal{K}$.

The story $\mathcal{S}$ is a (possibly empty) set of atomic sentences, each of which contains some $\kappa \in \mathcal{K}$ to which $\mathcal{I}$ has assigned no $u \in \mathcal{U}$; no atom having the distinguished unary symbol Ω as predicate may appear in $\mathcal{S}$.

The interpretation given to Ω is the entire universe $\mathcal{U}$; this predicate symbol is used in atoms like $\Omega(\kappa)$ or $\neg\Omega(\kappa)$ to talk about whether or not $\mathcal{I}$ maps $\kappa \in \mathcal{K}$ to some $u \in \mathcal{U}$; $\Omega(\kappa)$ can be read as "κ designates some element of $\mathcal{U}$" or as "the object denoted by κ exists" or simply as "κ exists."

The model-theoretic rules for determining truth in FOFL are given below. Except for rule (a), which specifies model theory for atomic sentences, these rules are the same as their counterparts in FOCL. There are two parts to rule (a): the first part specifies model theory for those atomic sentences where the predicate is the distinguished symbol Ω; the second part of the rule specifies model theory for atomic sentences not having Ω as predicate. The first part of the rule specifies that a sentence of the form $\Omega(\kappa)$ is true just in those cases where $\mathcal{I}$ does map κ onto something in the universe of discourse. The second part of the rule specifies that an atomic sentence whose predicate is not Ω can be true, not just in the usual way familiar from FOCL, but also if the sentence appears in $\mathcal{S}$ (which also means that the sentence must contain at least one constant symbol that is not mapped into $\mathcal{U}$).

In these rules, note that $\mathcal{M}_{\{X \mapsto u\}}$ denotes an extended version of the model $\mathcal{M} = \langle \mathcal{U}, \mathcal{I} \rangle$, in which the interpretation function $\mathcal{I}$ is extended to $\mathcal{I} \cup \{X \mapsto u\}$.

(a-1) $\mathcal{M} \models \Omega(\kappa)$ iff $\mathcal{I}(\kappa)$ is in $\mathcal{U}$.
(a-2) For all p in $\mathcal{P}$, $\mathcal{M} \models p(a_1, ..., a_n)$
iff $\langle \mathcal{I}(a_1), ..., \mathcal{I}(a_n) \rangle$ is in $\mathcal{I}(p)$ or $p(a_1, ..., a_n)$ is in $\mathcal{S}$.
(b) $\mathcal{M} \models \neg A$ iff $\mathcal{M} \not\models A$.
(c) $\mathcal{M} \models A \wedge B$ iff $\mathcal{M} \models A$ and $\mathcal{M} \models B$.
(d) $\mathcal{M} \models A \vee B$ iff $\mathcal{M} \models A$ or $\mathcal{M} \models B$.
(e) $\mathcal{M} \models A \Rightarrow B$ iff $\mathcal{M} \not\models A$ or $\mathcal{M} \models B$.
(f) $\mathcal{M} \models A \Leftrightarrow B$ iff ($\mathcal{M} \not\models A$ and $\mathcal{M} \not\models B$) or ($\mathcal{M} \models A$ and $\mathcal{M} \models B$).
(g) $\mathcal{M} \models (\forall X)A$ iff $\mathcal{M}_{\{X \mapsto u\}} \models A$ for every $u \in \mathcal{U}$.
(h) $\mathcal{M} \models (\exists X)A$ iff $\mathcal{M}_{\{X \mapsto u\}} \models A$ for some $u \in \mathcal{U}$.

4 Free Constraint Processing

Based on the relationship between classical logic and conventional constraint networks and on the relationship between classical and free logic, it is possible to define the notion of a free constraint network.

Free Constraint Network:
A free constraint network is a quadruple $\langle \mathcal{U}, \nabla, \mathbf{X}, \mathbf{C} \rangle$, where $\mathcal{U}$ is a universe of discourse, ∇ is a distinguished entity, $\nabla \notin \mathcal{U}$, $\mathbf{X}$ is a non-empty, possibly infinite, tuple of non-recurring parameters, and $\mathbf{C}$ is a finite set of r constraints. Each constraint $C_k(T_k) \in \mathbf{C}$, where T_k is a sub-tuple of $\mathbf{X}$, is a subset of the Cartesian product $(\Pi_{X_j \text{ in } T_k}(v(X_j)))$ where $v(X) = (\mathcal{U} \cup \{\nabla\})$.

Thus, each constraint $C_k(T_k) \in \mathbf{C}$ is a possibility distribution which restricts the existence of the parameters in T_k, as well as the values that these parameters may assume if they exist. $C_k(T_k)$ is a subset of a Cartesian product which contains one factor of the form $(\mathcal{U} \cup \{\nabla\})$ for each parameter in the tuple T_k. Consider, for example, the constraints

$$C_1(x, y) = \{\langle \nabla, 1 \rangle, \langle \nabla, 3 \rangle, \langle \nabla, \nabla \rangle\}$$
$$C_2(y, z) = \{\langle 1, 2 \rangle, \langle 2, 3 \rangle, \langle \nabla, 2 \rangle\}$$

in the network

$$\langle \mathcal{U}, \mathbf{X}, \mathbf{C} \rangle = \langle \{1, 2, 3\}, \nabla, \langle x, y, z \rangle, \{C_1(x, y), C_2(y, z)\} \rangle.$$

The constraint $C_1(x, y)$ is a strict subset of $(\mathcal{U} \cup \{\nabla\}) \times (\mathcal{U} \cup \{\nabla\})$. This constraint specifies that the parameter x cannot denote anything in the universe of discourse, because it only allows x to be mapped onto ∇; the constraint allows y either to be mapped onto ∇ or to denote something in $\mathcal{U}$, but it does specify that, if y does denote something in $\mathcal{U}$, then it must denote either 1 or 3. The second constraint $C_2(y, z)$ also allows y either to be mapped onto ∇ or to denote something in $\mathcal{U}$, but it specifies that, if y denotes something in $\mathcal{U}$, then it must denote either 1 or 2; this constraint specifies that z must denote something in the universe of discourse, since it requires that the parameter should be mapped onto either 2 or 3.

Free constraint networks are a generalization of conventional networks. A conventional constraint network is a free constraint network containing a finite number of parameters, all of which exist; that is, in a conventional network $\mathbf{X}$ is a finite tuple, and the constraint(s) in $\mathbf{C}$ unconditionally require(s) every parameter in $\mathbf{X}$ to denote something in $\mathcal{U}$. In contrast, a free constraint network may, in general, contain an infinite number of parameters, none of which need, in general, denote anything in the universe of discourse.

The intent of a free constraint network is a joint possibility distribution on the existence of the parameters in the network and on the values that these parameters may assume, if they exist:

Intent of a Free Constraint Network:
The intent of a free constraint network $\langle \mathcal{U}, \nabla, \mathbf{X}, \mathbf{C} \rangle$ is

$$\Phi_{\mathcal{U},\mathbf{X},\mathbf{C}} = \hat{C}_1(\mathbf{X}) \cap \ldots \cap \hat{C}_r(\mathbf{X}),$$

where $\hat{C}_j(\mathbf{X})$ is the cylindrical extension of the constraint $C_j(T_j)$ in the Cartesian product $(\Pi_{X_j \ in \ \mathbf{X}}(v(X_j)))$, the existence and value space for $\mathbf{X}$.

Consider, for example, the free constraint network given above. The cylindrical extensions of the constraints are:

$$\hat{C}_1(x, y, z) = \{\langle \nabla, 1\rangle, \langle \nabla, 3\rangle, \langle \nabla, \nabla\rangle\} \times \{\nabla, 1, 2, 3\}$$
$$\hat{C}_2(x, y, z) = \{\nabla, 1, 2, 3\} \times \{\langle 1, 2\rangle, \langle 2, 3\rangle, \langle \nabla, 2\rangle\}.$$

Intersecting these, we find that the intent of the network is

$$\Phi_{\mathcal{U},\mathbf{X},\mathbf{C}} = \{\langle \nabla, 1, 2\rangle, \langle \nabla, \nabla, 2\rangle\}.$$

The possibility that some parameters in a free constraint network need not exist (need not be mapped into $\mathcal{U}$) means that several additional forms of CSP may be defined [1, 2]. Here, however, we are interested in just one of these, the Minimal Exemplification CSP, which is defined below, after some subsidiary concepts have been defined.

Interpretation Set:
Given a mapping from a tuple of parameters T_k to a tuple of values t_k, the expression $\iota(T_k, t_k)$ denotes the corresponding interpretation set, the set of all those parameter to value mappings in which the parameters denote something in the universe $\mathcal{U}$.

For example, given $T_k = \langle y, z\rangle$ and $t_k = \langle 2, 3\rangle$, the expression $\iota(T_k, t_k)$ denotes the set of parameter to value mappings $\{y \mapsto 2,\ z \mapsto 3\}$. But, given $T_k = \langle y, z\rangle$ and $t_k = \langle \nabla, 2\rangle$, the expression $\iota(T_k, t_k)$ denotes the set $\{z \mapsto 2\}$, which contains only one mapping since y does not denote something in the universe of discourse.

Minimal Intent of a Free Constraint Network: The minimal intent of a free constraint network $\langle \mathcal{U}, \nabla, \mathbf{X}, \mathbf{C} \rangle$, written $\mu\Phi_{\mathcal{U},\mathbf{X},\mathbf{C}}$, is
$\mu\Phi_{\mathcal{U},\mathbf{X},\mathbf{C}} = \{Y | Y \in \Phi_{\mathcal{U},\mathbf{X},\mathbf{C}} \wedge \neg((\exists Z)\ (Z \in \Phi_{\mathcal{U},\mathbf{X},\mathbf{C}} \wedge \iota(\mathbf{X}, Z) \subset \iota(\mathbf{X}, Y)))\}$.

For example, the minimal intent of the above free constraint network is $\mu\Phi_{\mathcal{U},\mathbf{X},\mathbf{C}} = \{\langle\nabla,\nabla,2\rangle\}$. The tuple $\langle\nabla,1,2\rangle$, which was in the regular intent of the network, is not in the minimal intent since $\iota(\langle x,y,z\rangle,\langle\nabla,\nabla,2\rangle) \subset \iota(\langle x,y,z\rangle,\langle\nabla,1,2\rangle)$.

Minimal Exemplification CSP: Given a free constraint network $\langle\mathcal{U},\nabla,\mathbf{X},\mathbf{C}\rangle$, return some tuple in $\mu\Phi_{\mathcal{U},\mathbf{X},\mathbf{C}}$, if $\Phi_{\mathcal{U},\mathbf{X},\mathbf{C}}$ is non-empty; otherwise return nil.

Thus, solving the Minimal Exemplification CSP involves determining whether there can be any valuation for the parameters in the network such that this valuation satisfies all the constraints; in addition, if there can be such a valuation, solving the Minimal Exemplification CSP involves finding such one such valuation which has the property that the constraints would not be satisfied if any one of the parameters which is mapped by the valuation onto an element of the universe of discourse were instead mapped onto ∇.

5 Free Constraint Networks and Free Logic

A free constraint network $\langle\mathcal{U},\nabla,\mathbf{X},\mathbf{C}\rangle$ specifies the possibility distribution for the existence, as well as the interpretation, of constant symbols which appear in a free logic theory Γ, written in a first-order free language $\mathcal{L} = \langle\Omega,\mathcal{P},\mathcal{F},\mathcal{K}\rangle$. As in the conventional case, the definition of the network depends on Γ, the universe of discourse $\mathcal{U}$, and a partial interpretation function $\mathcal{I}_p$ for $\mathcal{L}$ which gives interpretations for all function and predicate symbols, and some of the constant symbols, of $\mathcal{L}$. However, the free network definition also depends on the story $\mathcal{S}$.

Taking any tuple Y in $\Phi_{\mathcal{U},\mathbf{X},\mathbf{C}}$ and computing $\mathcal{I} = \mathcal{I}_p \cup \iota(\mathbf{X}, Y)$ produces an interpretation function $\mathcal{I}$ such that the theory Γ is satisfied under the free logic model $\mathcal{M} = \langle\mathcal{U},\mathcal{I},\mathcal{S}\rangle$ of $\mathcal{L}$. Similarly, taking any tuple Z in $\mu\Phi_{\mathcal{U},\mathbf{X},\mathbf{C}}$ and computing $\mathcal{I} = \mathcal{I}_p \cup \iota(\mathbf{X}, Z)$ produces a minimal interpretation function $\mathcal{I}$ such that the theory Γ is satisfied under the minimal free logic model $\mathcal{M} = \langle\mathcal{U},\mathcal{I},\mathcal{S}\rangle$ of $\mathcal{L}$.

5.1 Example 4

Consider, for example, the following situation:

Language: $\mathcal{L} = \langle\Omega,\{p,q\},\{\},\{1,2,3,a,b,c\}\rangle$
Theory: $\Gamma = \{p(a,b), p(a,c) \Rightarrow q(a,c)\}$
Universe of discourse: $\mathcal{U} = \{1,2,3\}$
Story: $\mathcal{S} = \{p(a,b)\}$
Partial Interpretation:

$$\mathcal{I}_p = \{1 \mapsto 1, 2 \mapsto 2, 3 \mapsto 3, \Omega \mapsto \{1,2,3\},$$
$$p \mapsto \{\langle 1,2\rangle,\langle 2,3\rangle,\langle 3,3\ rangle\}, q \mapsto \{\langle 1,2\rangle,\langle 2,3\rangle\}\}.$$

The free constraint network $\langle\mathcal{U},\nabla,\mathbf{X},\mathbf{C}\rangle$ corresponding to this is the possibility distribution for the existence and interpretation of a, b and c, the three constant symbols which are not interpreted by $\mathcal{I}_p$. The components $\mathcal{U}$, $\mathbf{X}$ and $\mathbf{C}$ of the network are as follows: $\mathcal{U} = \{\{1,2,3\}\}$; $\mathbf{X} = \langle a,b,c\rangle$; $\mathbf{C} = \{C_1(a,b), C_2(a,c)\}$.

$C_1(a,b)$ corresponds to $p(a,b) \in \Gamma$. Since $p(a,b) \in \mathcal{S}$, at least one of a or b must not exist. Thus, $C_1(a,b) = (\{\nabla\} \times \{\nabla,1,2,3\}) \cup (\{\nabla,1,2,3\} \times \{\nabla\})$. $C_2(a,c)$ corresponds to $p(a,c) \Rightarrow q(a,c)$. Since this can be rewritten as $\neg p(a,c) \vee q(a,c)$, $C_2(a,c)$ can be defined as the union of two sets, one for each disjunct. Since $p(a,c) \notin \mathcal{S}$, $p(a,c)$ is false if a does not exist or if c does not exist or if a and c both exist but do not satisfy p. Since $q(a,c) \notin \mathcal{S}$, $q(a,c)$ is true iff a and c both exist and satisfy q. Calculating the union of the corresponding possibility distributions, and simplifying, we get $C_2(a,c) = (\{\nabla,1,2,3\} \times \{\nabla,1,2,3\}) - \{\langle 3,3\rangle\}$. Intersecting the cylindrical extensions of these constraints, we get $\Phi_{\mathcal{U},\mathbf{X},\mathbf{C}} = (((\{\nabla\} \times \{\nabla,1,2,3\}) \cup (\{\nabla,1,2,3\} \times \{\nabla\})) \times \{\nabla,1,2,3\}) - \langle 3,\nabla,3\rangle\}$, which means that $\mu\Phi_{\mathcal{U},\mathbf{X},\mathbf{C}} = \{\langle\nabla,\nabla,\nabla\rangle\}$. This means that there are 27 different models $\mathcal{M} = \langle\mathcal{U},\mathcal{I},\mathcal{S}\rangle, \mathcal{I}_p \subseteq \mathcal{I}$, of the language $\mathcal{L}$ under which Γ is satisfied; there is only one minimal model, $\langle\mathcal{U},\mathcal{I}_p,\mathcal{S}\rangle$, and under this model neither a, b nor c exist.

5.2 Example 5

To see the impact of the story on the intent of a theory, consider the above situation, modified so that the story $\mathcal{S} = \{\}$. The only difference in the network is that $C_1(a,b) = \{\langle 1,2\rangle,\langle 2,3\rangle,\langle 3,3\rangle\}$. However, computing the intent and the minimal intent, we get $\Phi_{\mathcal{U},\mathbf{X},\mathbf{C}} = (\{\langle 1,2\rangle,\langle 2,3\rangle,\langle 3,3\rangle\}\ times\{\nabla,1,2,3\}) - \{\langle 3,3,3\rangle\}$, and $\mu\Phi_{\mathcal{U},\mathbf{X},\mathbf{C}} = \{\langle 1,2,\nabla\rangle,\langle 2,3,\nabla\rangle,\langle 3,3,\nabla\rangle\}$, so, although c need not exist, both a and b must exist when the story is empty.

6 Galileo-2 Programs

Based on the above ideas, we have implemented a constraint programming language called Galileo-2. A program in Galileo-2 is a declarative specification of a free constraint network, analogous to the problem specification in Example 4. That is, in general, a Galileo-2 program specifies a first-order language $\mathcal{L} = \langle\Omega,\mathcal{P},\mathcal{F},\mathcal{K}\rangle$, a theory Γ containing sentences from that language, a universe of discourse $\mathcal{U}$, a partial interpretation function $\mathcal{I}_p$ for $\mathcal{L}$, and a story $\mathcal{S}$. Of these, only the theory Γ must always be specified explicitly. The Galileo-2 run-time system provides a default language $\mathcal{L}_g = \langle\Omega,\mathcal{P}_g,\mathcal{F}_g,\mathcal{R}\rangle$ in which $\mathcal{R}$ contains the real numeric strings, $\mathcal{P}_g$ contains names of standard predicates (=, =<, etc.), and $\mathcal{F}_g$ contains names of standard functions (*, +, etc.). The run-time system also provides a universe of discourse $\mathcal{U}_g = \Re$, an interpretation function $\mathcal{I}_g$ for $\mathcal{L}_g$ in terms of $\mathcal{U}_g$ and a story $\mathcal{S}_g = \{\}$.

The language $\mathcal{L} = \langle\Omega,\mathcal{P},\mathcal{F},\mathcal{K}\rangle$ defined by a Galileo-2 program has the following components: $\mathcal{P} = \mathcal{P}_g \cup$ {predicate symbols defined in the program}; $\mathcal{F} = \mathcal{F}_g \cup$ {function symbols defined in the program}; $\mathcal{K} = \mathcal{R} \cup$ {constant symbols used in the program}. The universe of discourse $\mathcal{U}$ defined by a Galileo-2 program is the union of $\mathcal{U}_g = \Re$ with any application-specific domains that are defined in the program. The partial interpretation function $\mathcal{I}_p$ defined by the program is the union of the set of mappings in $\mathcal{I}_g$ with the mappings provided by any definitions of application-specific domains, relations and functions that are in the program. The story $\mathcal{S}$ defined by a Galileo-2 program is the union of $\mathcal{S}_g = \{\}$ with any story provided in the program.

6.1 Program 1

The Galileo-2 program corresponding to Example 1 above is as follows:

```
area : positive.
velocity : nonnegative.
volume : nonnegative.
volume = area * velocity.
velocity =< 200.
```

A Galileo-2 statement of the form "$\kappa : p$" is merely an elliptical form of the Galileo-2 statement "exists(κ) and $p(\kappa)$," where "exists(κ)" is Galileo-2 syntax for $\Omega(\kappa)$.

The language $\mathcal{L} = \langle \Omega, \mathcal{P}, \mathcal{F}, \mathcal{K} \rangle$ defined by this program has the following components: $\mathcal{P} = \mathcal{P}_g$; $\mathcal{F} = \mathcal{F}_g$; $\mathcal{K} = \mathcal{R} \cup \{area, velocity, volume\}$. The free logic theory introduced by this program is

$$\{ \Omega(area) \wedge positive(area), \\ \Omega(velocity) \wedge nonnegative(velocity), \\ \Omega(volume) \wedge nonnegative(volume), \\ volume = area * velocity, \\ velocity \leq 200 \}.$$

The universe of discourse $\mathcal{U} = \mathcal{U}_g = \Re$. The partial interpretation function $\mathcal{I}_p = \mathcal{I}_g$. The story $\mathcal{S} = \mathcal{S}_g = \{\}$. This program, in effect, defines a conventional constraint network because the first conjunct in the first three sentences of the theory specify that the objects denoted by the constant symbols *area*, *velocity* and *volume* must exist.

6.2 Program 2

However, the following Galileo-2 program, which corresponds to Example 3, does define a free constraint network, because the existence of *radius* or of *breadth* and *height* is contingent on the interpretation of *shape*:

```
domain form =::= {circular, rectangular}.
relation standard =::= {46.9,57.8,95.7,124.7}.
area : positive.
velocity : nonnegative.
volume : nonnegative.
shape : form.
volume = area * velocity.
velocity =< 200.
exists (X,Y) : standard(X) and standard(Y) and
                    not(X=Y) and area < X and area < Y.
shape=circular implies
            exists(radius) and positive(radius) and area = 3.14159 * radius^ 2.
shape=rectangular implies
            exists(breadth) and positive(breadth) and
            exists(height) and positive(height)
            and area = breadth * height.
exists(radius) equiv not exists(breadth) and not exists(height).
exists(height) equiv exists(breadth).
```

In reading this program, it is important to distinguish between quantification and usage of the special existence predicate. In Galileo-2, logic variables can be distinguished from other symbols because they are tokens whose first character is a capital letter. Quantification is written as follows: "$(\forall X)(p(X))$" is written as "all X : p(X)," "$(\exists X)(p(X))$" is written as "exists X : p(X)" and, when several variables are subject to the same type of quantifier, they can be grouped together; thus, $(\exists X)(\exists Y)(p(X, Y))$ is written as "exists (X,Y) : p(X,Y)." A Galileo-2 phrase of the form "exists(κ)" where κ is a constant symbol is not an instance of quantification; it involves usage of the free logic existence predicate.

The language $\mathcal{L} = \langle \Omega, \mathcal{P}, \mathcal{F}, \mathcal{K} \rangle$ defined by this program has the following components: $\mathcal{P} = \mathcal{P}_g \cup \{form, standard\}$; $\mathcal{F} = \mathcal{F}_g$; $\mathcal{K} = \mathcal{R} \cup \{circular, rectangular, area, velocity, volume, shape, radius, breadth, height\}$.

The theory introduced by this program is

$$
\begin{aligned}
\{ & \Omega(area) \wedge positive(area), \\
& \Omega(velocity) \wedge nonnegative(velocity), \\
& \Omega(volume) \wedge nonnegative(volume), \\
& \Omega(shape) \wedge form(shape), \\
& volume = area * velocity, \quad velocity \leq 200, \\
& (\exists X)(\exists Y)(standard(X) \wedge standard(Y) \wedge \neg(X = Y) \wedge area < X \wedge area < Y), \\
& shape = circular \Rightarrow \Omega(radius) \wedge positive(radius) \wedge area = 3.14159 * radius^2, \\
& shape = rectangular \Rightarrow \\
& \qquad \Omega(breadth) \wedge positive(breadth) \wedge \\
& \qquad \Omega(height) \wedge positive(height) \wedge \\
& \qquad area = breadth * height, \\
& \Omega(radius) \Leftrightarrow \neg\Omega(breadth) \wedge \neg\Omega(height), \\
& \Omega(height) \Leftrightarrow \Omega(breadth)\}.
\end{aligned}
$$

The universe of discourse is $\mathcal{U} = \mathcal{U}_g \cup$ {`circular`,`rectangular`} $= \Re \cup$ {`circular`,`rectangular`} The partial interpretation function $\mathcal{I}_p = \mathcal{I}_g \cup \{$ *form* $\mapsto$ {`circular`, `rectangular`}, *standard* $\mapsto$ {`46.9`,`57.8`,`95.7`,`124.7`}, *circular* $\mapsto$ `circular`, *rectangular* $\mapsto$ `rectangular` $\}$. The story $\mathcal{S} = \mathcal{S}_g = \{\}$.

The two implication statements in this program specify when *radius*, *breadth* and *height* <u>must</u> exist, i.e., these statements specify when the parameters must denote entities in the universe of discourse. These statements do not specify when the parameters <u>can</u> exist, i.e., the statements do not specify when the parameters are allowed to denote entities in the universe of discourse. This is accomplished by the addition of the two equivalence statements. The cardinality of the network intent would be greatly increased by the elimination of the equivalence statements. If these statements were eliminated, the resultant network would be satisfied if, for example, *breadth* and *height* existed even in situations where the shape was circular; thus, tuples representing this kind of situation would also appear in the network intent. However, although the regular intent of the network would be expanded by the elimination of the equivalence statements, the minimal intent of the network would be unchanged: the minimal intent only contains those tuples which represent situations in which the only parameters that denote entities of the universe of discourse are exactly those parameters which are required to do so.

6.3 Program 3

The following program, which corresponds to Example 4, contains two application-specific relation definitions, as well as an application-specific story:

```
relation p =::= {(1,2), (2,3), (3,3)}.
relation q =::= {(1,2), (2,3)}.
story =::= {p(a,b)}.
p(a,b).
p(a,c) implies q(a,c).
```

The language $\mathcal{L} = \langle \Omega, \mathcal{P}, \mathcal{F}, \mathcal{K} \rangle$ defined by this program has the following components: $\mathcal{P} = \mathcal{P}_g \cup \{p, q\}$; $\mathcal{F} = \mathcal{F}_g$; $\mathcal{K} = \mathcal{R} \cup \{a, b, c\}$. The theory introduced by this program is $\{p(a, b),$

$p(a,c) \Rightarrow q(a,c)\}$. The universe of discourse is $\mathcal{U} = \mathcal{U}_g = \Re$. The partial interpretation function $\mathcal{I}_p = \mathcal{I}_g \cup \{ p \mapsto \{ \langle 1,2\rangle, \langle 2,3\rangle, \langle 3,3\rangle \}, q \mapsto \{ \langle 1,2\rangle, \langle 2,3\rangle \} \}$. The story $\mathcal{S} = \mathcal{S}_g \cup \{p(a,b)\} = \{p(a,b)\}$. The language $\mathcal{L}$ defined by this program is larger than the language in Example 4, because of the presence of the default language $\mathcal{L}_g$ provided by the Galileo-2 run-time system; however, the free constraint network defined by this program is identical to that in Example 4.

6.4 Galileo-2 and Full Free Logic

A complete description of the Galileo-2 language is beyond the scope of this paper. However, it should be noted that the language supports the full FOFL. Thus, for example, Galileo-2 supports programs that define theories which contain arbitrarily nested quantified sentences. In addition, although none of the programs considered here involved application-specific functions, the definition and use of such functions is supported.

7 The Galileo-2 Run-time System

Since FOFL subsumes FOCL, satisfiability for FOFL is undecidable. There cannot exist, therefore, an algorithm capable of performing automatic constraint satisfaction on an arbitrary network specified in Galileo-2. However, the Galileo-2 run-time system, which can be run in either autonomous or interactive mode, can solve various forms of CSP for a wide variety of networks. Here, we focus on the Minimal Exemplification CSP. The Galileo-2 run-time system, in autonomous mode, is capable of autonomously solving the Minimal Exemplification CSP for those free networks in which all parameters have finite domains and also for those networks in which some or all parameters have infinite domains, provided the networks are decidable by backtrack-free search. In interactive mode, the run-time system can solve the Minimal Exemplification CSP for any network which the user, by non-monotonically augmenting the theory with additional assertions, renders decidable by backtrack-free search. Since any model of an augmented theory is also a model of the original theory, any non-nil solution to the Minimal Exemplification CSP for the augmented network is a solution for the original network.

Full details of constraint processing in Galileo-2 are beyond the scope of this paper. However, an inference algorithm called Compound Constraint Propagation [5] is a central part of both the backtracking searcher and the non-backtracking searcher. Its top-level is as follows:

```
procedure CCP(Assertions)
localvar Q_lpks, Q_gac, Q_gpc
begin
     Q_lpks ← Assertions; Q_gac ← {}; Q_gpc ← {};
     repeat
       LPKS(Q_lpks, Q_gac, Q_gpc);
       if Q_gac ≠ {} ∨ Q_gpc ≠ {}
          then repeat
                  if Q_gac ≠ {}
                       then GAC(Q_lpks, Q_gac, Q_gpc);
                  if Q_lpks = {} ∧ Q_gpc ≠ {}
                       then GPC(Q_gac, Q_gpc)
               until Q_lpks ≠ {} ∨ Q_gac = {}
     until Q_lpks = {};
end
```

The algorithm involves interleaved application of three inference techniques:

- a version of local propagation of known states, extended to assimilate conditional existence; this is performed by invoking a lower-level procedure called LPKS;
- a version of arc consistency, generalized to infinite domains and constraints of arbitrary arity; this is performed by invoking procedure GAC;
- a form of path consistency, generalized to infinite domains and constraints of arbitrary arity; this operation, which is performed by invoking procedure GPC, is only applied to small portions of the network in certain very specific circumstances.

8 Comparative Discussion

By basing our approach to constraint networks on free logic, we have been able to give our work what seems to be a more developed theoretical underpinning than the only other known approach to conditional existence of network parameters, that of Mittal and Falkenhainer [11]. In addition, our approach is more general. Their notion of a dynamic constraint network is a highly restricted special case of our concept of a free constraint network.

In [11], the distinction between "active" and "inactive" parameters refers to whether the parameters denote entities in the universe of discourse. In this paper, we have concentrated on this distinction too. In addition, however, the Galileo-2 run-time system uses the distinction between denotation and non-denotation to enable a further distinction, that between parameters which exist in the name space of the run-time system and those which do not. Although an illustration and explanation are beyond the scope of this paper, this second distinction has an important pragmatic consequence: it means that our approach can handle problems with an infinite number of potentially-existent parameters, because storage space is not used by network parameters until such time as their actual existence is proven or disproven. Indeed, in most cases even the *names* of potentially-existent parameters are generated only when the parameters' existence is proven. In practical applications with an infinity of potentially-existent parameters, this proof of existence happens for only a finite subset of the infinity of parameters. Therefore, no storage burden is imposed by the infinity of other parameters whose possible existence is irrelevant to the particular minimal model constructed. In Galileo-2, the names of the infinity of potentially-existent parameters are treated as functional expressions in a meta-theory. However, the use of meta-theory in Galileo-2 is beyond the scope of this paper.

Free logic is a principled attempt to remedy an ontological inadequacy of classical logic. This logic has long been studied by philosophers and has recently [8] attracted some attention in the Knowledge Representation literature. Nevertheless, although there are several computer languages which are explicitly based on first-order classical logic, our language Galileo-2, and its predecessor Galileo, seem to be the only languages based on free logic.

References

[1] Bowen J, and Bahler D, 1990, "Improving Ontological Expressiveness in Constraint Processing," Technical Report, Department of Computer Science, North Carolina State University.

[2] Bowen J, and Bahler D, 1991, "Free Logic in Constraint Processing," Technical Report, Department of Computer Science, North Carolina State University. Submitted to *Artificial Intelligence* journal.

[3] Bowen J, and Bahler D, 1991, "Conditional Existence of Variables in Generalized Constraint Networks," in *Proceedings of AAAI-91, the National Conference of the American Association for Artificial Intelligence*, Anaheim CA, July 1991.

[4] Bowen J, O'Grady P and Smith L, 1990, "A Constraint Programming Language for Life-Cycle Engineering," *International Journal for Artificial Intelligence in Engineering*, 5(4), 206-220.

[5] Bowen J, and Bahler D, 1992, "Compound Constraint Propagation," Technical Report, Department of Computer Science, North Carolina State University.

[6] Bowen J, Bahler D, and Dholakia A, 1990, "A DFT Advisor for Digital Circuit Design," Technical Report, Department of Computer Science, North Carolina State University. To appear in *Computers and Electrical Engineering*, special issue on Artificial Intelligence in Engineering Design and Manufacturing.

[7] Friedman G and Leondes C, 1969, "Constraint Theory, Part I: Fundamentals," *IEEE Transactions on Systems Science and Cybernetics*, ssc-5, 1, 48-56.

[8] Hirst G, 1989, "Ontological assumptions in knowledge representation," *Proceedings of the First International Conference on Principles of Knowledge Representation and Reasoning*, 157-169.

[9] Lambert K and van Fraassen B, 1972, *Derivation and Counterexample: An Introduction to Philosophical Logic*, Enrico, CA: Dickenson Publishing Company.

[10] Mackworth A, 1987. "Constraint Satisfaction," in S. Shapiro (ed.), *The Encyclopedia of Artificial Intelligence*, New York: Wiley, 205-211.

[11] Mittal S and Falkenhainer B, 1990, "Dynamic Constraint Satisfaction Problems," *Proceedings of the Eighth National Conference on Artificial Intelligence*, 25-32.

[12] Mulder J, Mackworth A and Havens W, 1988, "Knowledge Structuring and Constraint Satisfaction: The Mapsee Approach," *IEEE Transactions on Pattern Analysis and Machine Intelligence*, 10(6), 866-879.

[13] Nilsson N, 1991, "Logic and Artificial Intelligence," *Artificial Intelligence*, 47, 31-56.

Section 6:

Vision

Epistemics

Mental-Model and Mental-Logic based Reasoning about Space

Enrico Giunchiglia
DIST, University of Genoa
Genoa, Italy

Alessandro Armando
DIST, University of Genoa
Genoa, Italy

Abstract

The aim of this work is to describe (part of) a system able to visualize natural language descriptions of complex scenes. Indefinitely many objects may take part in the scenario and are introduced one at any time with their relative positions (expressed as a certain number of spatial predicates). The specific goal of this paper is to specify which reasoning needs to be done between the input natural language descriptions and the visualization of the scene. The key idea of our approach is to use both a declarative representation of space based on first order logic, and a procedural one based on a semantic network. The two representations are linked defining the semantic network to be the *intended model* of the first order theory. The first order and the semantic network based descriptions can be respectively regarded as the system *mental-logic model* and *mental model* of the scenario.

1 Introduction

In the last few years, CAD systems have been evolving from simple drafting tools to much more complex solid modelling environments. Features of such systems are usually the possibility of performing orthogonal and/or perspective projections, geometrical operations such as sections and mechanical features computation (e.g. inertia moment), and so on. On the other hand, if we think of systems for scene generation, they usually only offer a set of basic geometric operations which the user can compose to obtain his scope. The key point is: how are such basic geometric primitives "user-friendly"? Is it always necessary to use such hyper-grained primitives? If we think about drawing a picture with a pen on a table, using the usual geometric primitives we have to specify in which "exact" position it is and which is its orientation (supposing we already have a geometric model for the pen). Such level of detail, though necessary in general, is not always not necessary for instance when the exact position and orientation is not relevant but what matters is that the pen is on the table. Thinking about the operation the system has to perform if told "put a pen on the table", we want him to be able to "reason" in the most "human-like" way.

The aim of this paper is to give some leading ideas and results which seem useful in order to build a suitable representation of space for systems able to reason in a "human-like" way. The key idea of the work described in this paper is to have a first order "qualitative" description as well as a semantic network [2] containing a more extensive "quantitative" description of the same space. Given the input phrase, the system tries to disambiguate it by building the first order representation. For example, given the phrase the results of the analysis will be H_CONTACT(pen1,table2), saying the *pen number 1* is *horizontally supported* by the *table number 2* [1]. Such analysis is necessary since the same english preposition can be used in an another phrase but with a completely different spatial meaning (eg. *"The picture on the wall"*). At this step, the system only qualitatively disambiguates the input phrase. In fact it does not compute the *effective* position of the pen on the table. We say the system builds its *"mental logic"* image of the scenario. Such image is then further analyzed to effectively understand whether it may corresponds to a real scenario. This step is performed by giving spatial meaning to both the object and the preposition and trying to visualize the resulting scene on a screen. If this step succeeds, we say the system builds its own *"mental model"* [8] of the scenario. The two worlds are related defining the semantic network to be the model of the first order theory: the first order objects and relations are linked to the objects and relations of the semantic network by means of semantic attachment [15, 13].

We are not faced with the philosophical reasons of our approach or of others; this topic has been already faced in [8, 4, 7]. Neither are we faced with the aim of the complete work and its possible applications (see [1, 6]). The reader is recommended to read those papers to get a complete view of the historical development and related work.

The paper follows this path: in the next section we present some leading ideas which have driven the system design and implementation. In section three we show how it is possible to reason about a scene starting from a human-like ambiguous description. As we will see, the result will be the construction of a first order theory (the system *mental-logic model* of the scenario) representing a first "qualitative" description. In the fourth section we focus on the mechanisms enabling the system to build a semantic network based description of the scenario. Such a description *completely* characterize the system *mental model* [8] of the scenario. Based only on the information stored in the semantic network, the system visualize *its mental model* on a screen. Finally in the last section it is shown how the the two representations can be linked via "semantic attachment" [13, 15].

2 Integrating qualititative and quantitative reasoning about space

Consider the task of understanding completely (ie. make a picture) a scene described by a natural language description. Supposing you are said

"There is a pen on the right of a bottle"

You probably make a first initial model figuring out there must be a table (since you are not used to put bottles and pens on the floor). Moreover supposing you have inferred (due to previous phrases) there are more than one table and/or pens and/or bottles, you question about *which* istances the user refers to. Only

[1]To simplify notation and comprehension, in the following we specify the instance numbers of objects only when really necessary.

as a final step you begin enriching/redrawing (supposing it is the case) your picture in such a way to satisfy the last phrase. During this last step, you have to *effectively* look at your picture to see whether your previous symbolic considerations are compatible with your already drawed picture. For example you have to check that object do not intersect and that both the pen and the bottle are in a stable position on the table.

In such a process we can thus recognize a first qualitative step of general evaluation of the problem and a final quantitative step of numerical values handling. Moreover, even though we do not recognize as distinct these two steps, our claim is that in order to build a system with human-like reasoning capabilities, a purely quantitative/not-structured approach would not have been sufficiently efficient and powerful.

In fact, as all the literature in qualitative reasoning points out, (for instance [5, 3, 9]) a first qualitative approach to the problem has the following advantages:

- it allows to reach conclusions with very little information available and, consequently, hard to formalize with quantitative models;
- qualitative reasoning can be very effective to reach approximate conclusions, sufficient in everyday life, at the right level of detail.

On the other hand logic allows a well formed definition of the problem and the possibility of moving within an extensively studied and well known framework. We have clear notion of wellformedness and deduction. We can say precisely which facts the conclusion we derive depends upon. We may not use some rules/axioms (because not adequate for the case) simply by not considering them during the deduction (not giving them as *premises* to the theorem prover). From an implementational point of view, we can change our "reasoning rules" (the axioms), affecting in such a way the system behavior, without modifying the system implementation.

Numerical computation is necessary to effectively check whether or not the qualitative description is compatible with the already derived quantitative description. More specifically, the system switches to the effective representation whenever the "qualitative reasoning" does not suffice. The link between the two worlds is made by a sort of "semantic attachment" [15, 13] which allows to associate a function to each predicative or functional letter and an object to each individual constant of the first order theory. By evaluating the resulting expression the system check the correspondence between the two descriptions.

To summarize, we have two distinct (but interacting) theories each one modeling space at a different level of detail:

- A first order theory of space written in some logic language, namely PROLOG. At this level deduction is performed both syntactically (as in any "classical" proof checker) and through semantic evaluation of functions and predicates (performing the "semantic attachment" informally stated in the previous section). This topic is described in detail in the next section.
- A semantic network where space is represented as a graph where the nodes are the objects in the scene and the links are labeled by the (natural language) spatial relations holding between them. In the semantic

network information is retrieved by procedural evaluation of the "spatial meanings" associated to the objects (the nodes of the semantic network) and prepositions (the links). Since during such evaluation, a lot of numerical computation has to be done (for example think when we have to check two objects do not intersect) this module has been written in the C-language.

The procedural evaluation is activated by a *deduction supervisor* which knows all the pairs < *syntactic object, intended meaning* > (the intended meaning of a syntactic object is a member of the domain D (of the intended model [10]) if the object is a constant, and a procedure, if the object is a functional or predicative symbol). Evaluating a syntactic expression corresponds to the activation of the associated (semantic) procedure given the (semantic) meaning of its subexpressions. As in [15, 13], the equality between the syntactic object whose intended meaning is the computed result and the expression being evaluated, is then asserted in the first order, ground theory. To simplify, consider the following example by arithmetic. Attach the numbers "2,3,5" respectively to the individual constants "two, three, five" and the procedure computing the sum of two numbers to the function symbol "plus". Evaluating "plus(two,three)" causes the following step to be performed:

1. the constant "two" is interpreted as the number "2". The same for the constant "three".
2. The function symbol "plus" is interpreted as the operation "+".
3. the expression built by substituting symbols in "plus(two,three)" with their interpretations (ie. "+(2,3)") is evaluated.
4. The system looks for an individual constants whose interpretation is equal to the computed result (ie. "5") (in our example, the constant "five").
5. The formula "plus(two,three) = five" is asserted in the first order theory.

Suppose we had no attachment to the number "5", (all the rest being equal to the former example), evaluating "plus(two,three)" would cause no assertion in the theory, (the system is not able to associate any first order constant to the computed number "5").

3 Building a *mental-logic model* of space

As briefly sketched in the introduction, the *mental-logic* model corresponds to a set of first order sentences. Considering the input phrase *"The pen on the table"* H_CONTACT(pen,table) is first derived via reasoning/deduction in the mental-logic. Such initial step corresponds to the disambiguation of the input relation, that is to the association of a predicative letter with an unambiguous spatial meaning to the input preposition.

Here, default reasoning [14] (expressed by considering some rules instead of others) plays an important role [1]. For example we derive that the preposition "on" means *horizontally on* since we have the default knowledge that tables are usually placed on the floor. But this is no longer true if in our environment we have a table being sticked to a wall. Thus, the conclusion we derive strictly relies on two different kinds of information:

- the a_priori knowledge of the world (this problem is not faced here, see [1])
- the current state of the world, which we *exactly* know via the semantic network representation.

To specify how default are handled, consider a *normal default* [2]. While we check the validity of prerequisites testing the knowledge base, the consistency of the conclusion with the real, present world is evaluated testing the semantic network. It tests the compatibility of what has been inferred previously with the output of the knowledge base and decides if backtracking has to be activated when exceptions arise.

Note that these rules are only an attempt to formalize (that is to give a cognitive model of) common sense reasoning about space. They do not take in consideration all the possible combinatorial situations. In fact some of them are cognitively impossible and are automatically excluded by the system.

An interesting example is:

$$\begin{array}{ll} RIGHT(pen, bottle) & \longrightarrow \\ H\_CONTACT(pen, table) & \wedge \\ H\_CONTACT(bottle, shelf) & \wedge \\ H\_CONTACT(table, floor) & \wedge \\ H\_CONTACT(shelf, floor) & \end{array}$$

As it can be seen we first deduce that the "pen" is horizontally supported by the table and then we recursively deduce the positions of all the inferred objects till the border of the environment [1, 7]. The rule applied when deriving H_CONTACT("pen", table) is a simplified form of the *"rule of the independent typical positions"* [1]. Formally, it can be described as in figure 1.

Note that we have supposed that the nearness of the two typical positions holds. Of course this is not always true, the contrary may happen even in this case, depending on how the table and the shelf are positioned. In this case the analysis changes: we position the object in its position but we do not know anything about the subject position. This case is taken into account in figure 2. The rule in figure 2 is part of the rule of the *"rule of the unknown subject position"*.

As a second step, the system tries to deduce more information about the predicates derived in the previous step. For instance it is obvious that if RIGHT(a,b) holds also LEFT(b,a) holds. This part of the theory is completely independent of the previous and can be seen as a first order description of the relations which exist among object positions. Some examples of the rules which applied are:

$$\forall x.\forall y.(RIGHT(x,y) \rightarrow (LEFT(y,x))$$
$$\forall x.\forall y.\forall z.(H\_CONTACT(x,y) \rightarrow \neg H\_CONTACT(x,z))$$

[2]Normal default looks like:

$$\frac{wff_1 : M wff_2}{wff_2}$$

wff_1 is said to be the *conclusion*, while wff_2 is said to be the *consequent* of the default [14].

$$W\_CONC(s,o)$$

/* the input conceptualization is a weak conceptualization */
/* (derived from on the right, in front of, near, ...). */

$$\wedge$$
$$TP(o)$$

/* the object has a typical position in the defined */
/* environment (evaluated intensionally on the a_priori */
/* data base) */

$$\wedge$$
$$NEAR\_POSIZ(s,o)$$

/* the subject and the object of the conceptualization are */
/* near to each other if positioned in their own typical */
/* positions (evaluated extensionally on a possible state */
/* of the semantic network) */

$$\wedge$$
$$(((TP\_H(s) \wedge TP\_H(o)) \vee (TP\_V(s) \wedge TP\_V(o)))$$

/* if positioned in their own typical position both the */
/* subject and the object are supported either by an */
/* horizontal or vertical surface (evaluated as TP(o)) */

$$\wedge$$
$$\neg COMMON\_MATRIX(s,o,m)$$

/* an object m which is the common matrix for the object */
/* and the subject does not exist. The common matrix of */
/* *n* objects is an object to which all the objects refer */
/* when positioned in their typical position */
/* (evaluated as TP(o)). */

$$\longrightarrow$$
$$TP\_CONC(s,s,kb\_tp\_obj(s)) \wedge TP\_CONC(o,o,kb\_tp\_obj(o))$$

/* both the subject and the object are positioned in their */
/* typical position, extracted from the data base a_priori */
/* with the function kb_tp_obj(o) */

Figure 1: *The rule of the independent typical positions.*

$$W\_CONC(s,o) \wedge TP(o)$$
$$\wedge$$
$$(((TP\_H(s) \wedge TP\_H(o)) \vee (TP\_V(s) \wedge TP\_V(o)))$$
$$\wedge$$
$$\neg NEAR\_POSIZ(s,o) \wedge \neg COMMON\_MATRIX(s,o,m)$$
$$\longrightarrow$$
$$TP\_CONC(o,o,kb\_tp\_obj(o)) \wedge IND\_POS(s)$$

/* The object is positioned in its typical position. */
/* As far as the subject is involved, we are not able to decide */
/* and ask to the user. */

Figure 2: *(Part of) of the rule of the unknown subject position.*

Of course, the theory we propose is largely incomplete. This is due to the extreme complexity of the problem which makes the problem unsolvable with this approach. To point out this fact it is sufficient to think to the infinite mutual positions that two objects may have. Moreover, as a consequence of the above consideration, in this step most reasoning is performed by evaluation of the property on the semantic network; only simple cases, such as those stated above, are treated symbolically. This topic is treated in the following sections.

4 Building a *mental model* of space

Consider the task of positioning a pen supposing you are told LEFT(pen,bottle). Without any doubt you could place the pen on the left of the bottle which is already on the table. What such a trivial example shows is that people seem able to infer the position of object not relatively to an absolute reference system, but to the objects the phrase refers to. In the case of the example, the agent is able to *correctly* place the pen, without knowing what is the exact position of the pen or of the bottle with respect of the environment.

Generalizing, when working about space representation we noted that, in an everyday discussion, object positions are nearly always defined relatively to the positions of other objects whose positions are recursively ill-defined. With such recursively ill-defined descriptions, humans are also able to check consistency. Thus, for instance, people are able to say that, if "pen1 is on the right of book2" and "pen3 is on the left of book2" then it is impossible that "pen1 is on the left of pen3". Moreover, when trying to position objects in a limited space people are also able to shift them maintaining all the known (ambiguous) constraints.

Such examples point out that it would be not appropriate to associate an absolute position to each object in the scenario. We consider a better solution (at least from a computational point of view) to associate to each (first order) relations among objects, a *compatibility function* specifying the allowed mutual positions. Such compatibility functions are the links of the semantic network, while the nodes are the objects the (first order) relation involves.

Actually, the compatibility functions take value in the range [0,1] and can be interpretd as fuzzy functions specifying the set of all the possible mutual object-subject positions with their compatibility/possibility/truth values. How

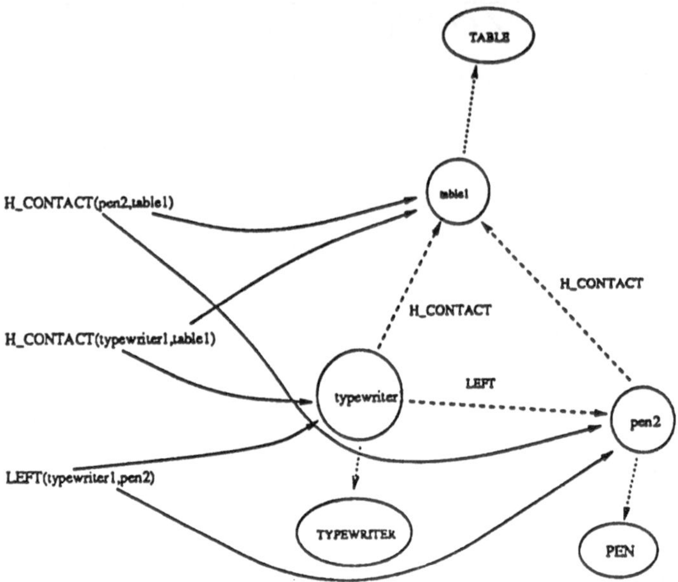

Figure 3: *An example of semantic network and the linking with the first order description.*

compatibility functions are built is largely explained in [6]; here only some notes are given. Taking the fuzzy approach we avoid all the on-off problems on the borders existing in the standard approaches [11, 12] and common sense reasoning can be performed in a more straightforward way.

For instance, we can interpretate the phrase "book on the table", as specifying not only that the book is horizontally supported by the table but also that, maybe, is in its center (or in a side ...) with an orientation which guarantees a high degree of equilibrium .... Of course such information has to be specified in the data-bases containing information about objects (i.e. the data-base must contain the fact that a book is usually in the center of the table while the phone is maybe on one side, more usually the right side). All the information about classes of objects is not duplicated for each object (node) in the semantic network, but to each object there is associated a pointer to the class of objects it belongs to. In figure 3 we see an example of a semantic network and the link with the first order description.

The functions which describe the mutual positions of the object and the subject of the spatial prepositions have some interesting features:

- first of all, they obviously depend on the given spatial preposition. Any spatial preposition refers to different object parts or characteristics (for instance **on** relates the volume of the subject to a surface of the object while **in** relates the two volumes);.
- The influence of an object on the function shape depends on its syntactic role in the spatial relationship (whether it is subject or object of the spatial relation) [6].

- Each function is built independently of any contextual check, the "contextual" space state is taken into account only when the overall "free space" compatibility function is synthesized and the relation (defined a_priori) between the above parameters and the function values, because of the very nature of common sense, is necessarily fuzzy and not crisp.

5 Linking the *mental logic* and the *mental model* representations

The link between the first order theory and the semantic network (implementing its intended model) is performed through a set procedures which are activated by the "deduction supervisor" which knows all the pairs:

$$< \textit{syntactic object}, \textit{intended meaning} >$$

Such procedures can effectively change the semantic network (and in this case also the scene on the screen will change) or only read the information in the semantic network to check the validity of a first order proposition. Notice that whenever we want to modify the semantic network or the first order knowledge database, we have to modify consistently the first order knowledge database or the semantic network. So far, any procedure affecting the state of the system, is responsible of maintaining a full compatibility between the theory and the semantic network. Of course most backtracking is activated at this level. In fact, when syntactically reasoning, the system activates a lot of default rules. This may be misleading and give some wrong deductions. Contradictions are most often discovered when checking the consistency of deductions with the knowldge in the semantic network. A discovered inconsistency, causes the retraction of one of the default which has been erroneously activated.

A complete description of the procedure evaluating or modifying the semantic network is beyond the aim of this work; see [6]. An example can be the analysis given in section 3. In that example the same input may give two different answers depending on the semantic evaluation of the predicate NEAR_POSIZ(s,o). This may be a good example of how semantics and syntactics cooperate and how the semantic attachment is performed. In fact the deduction supervisor, when working on NEAR_POSIZ, understands that the evaluation must be made procedurally and switches to the semantic level.

Finally, we justify our claim that the semantic network is the intended model of the first order description, noting that our representation of space completely fits with standard extensional semantics definition (as, for instance in [10]). In fact, consider the double valued version of our semantic network:

- D, (the domain of the interpretation), is taken to be $\Re^3$. It can be described, for example, by means of Cartesian coordinates.

- Any constant, (any instantiated object), is associated to an element belonging to D, that is a triple (x,y,z).

- Any variable, (any not instantiated object), is assumed to vary within D (we do not know where it is).

- Any function of arity n is assumed to get values in D^n and to take values in D.

- The standard logical operators are assumed to have their propositional value. So for, instance, the holding of a set of input phrases is interpreted as the set intersection of all the single sets;
- the predicates we have defined in our theory are interpreted as subsets of D. For instance, H_CONTACT(pen, table) is associated the set of all the points which are above the upper surface of the table.

6 Conclusions

We have tried to give only some general ideas of how we have modeled space, never explaining the details of the formalization. This has probably resulted in a sometimes not clear, always too brief and not precise, explanation. We hope that people reading this paper do not care of style and formal matters and understand the underlying ideas. The interested reader will find more a detailed description in [6].

References

[1] A. Adorni, M. DiManzo, and F. Giunchiglia. From descriptions to images: what reasoning in between? In *Proc. 6th European Conference on Artificial Intelligence*, September 1984.

[2] R.J. Brachman. What is-a is and isn't: An analysis of taxonomic links in semantic networks. *Computer IEEE*, October 1983.

[3] J.H. De Kleer and J.S. Brown. A qualitative physics based on confluences. *Artificial Intelligence.*, 24:7–83, 1984.

[4] M. DiManzo and F. Giunchiglia. A representation of space for knowledge based scene generation. In *Cognitiva 85*, June 1985.

[5] K.D. Forbus. Qualitative process theory. *Artificial Intelligence.*, 24:85–168, 1984.

[6] E. Giunchiglia, A. Armando, P. Traverso, and A. Cimatti. Reasoning about the configuration of spatial objects. Technical report, Dep. Comp. Science, University of Genoa, January 1991. Technical Report DIST-91-03.

[7] F. Giunchiglia, C. Ferrari, P. Traverso, and E. Trucco. Understanding scene descriptions by integrating different sources of knowledge. Technical report, Dep. Comp. Science, University of Genoa, October 1990. Technical Report DIST-90-54. To be published in the International Journal of Man Machines Studies.

[8] J. N. Johnson-Laird. *Mental Models.* Harvard University Press, 1983.

[9] B.J. Kuipers. Qualitative simulation. *Artificial Intelligence*, 29:289–338, 1986.

[10] Z. Manna. *Mathematical theory of computation.* Mc Graw hill Book Company, 1974.

[11] D. McDermott. Finding objects with given spatial properties. Technical Report Research Report n. 195, Dept. of computer Science, Yale University, 1981.

[12] D. McDermott and E. Davis. Planning routes through uncertain territory. *Artificial Intelligence*, Vol. 22:107–156, 1984.

[13] P. Pecchiari. Meccanizzazione del concetto di modello di un dimostratore interattivo. Master's thesis, University of Trento, Dept. of Mathematics, 1990. IRST-Thesis 9009-15.

[14] R. Reiter. A logic for default reasoning. *Artificial Intelligence*, 13, 1980.

[15] R.W. Weyhrauch. Prolegomena to a theory of mechanized formal reasoning. *Artificial Intelligence*, 13, 1980.

The Devil's Torpedo Tubes: a new Impossible Object considered in relation to the IO model of human vision.

Roddy Cowie
Rex Mitchell

School of Psychology, Queen's University of Belfast
Belfast BT7 1NN, United Kingdom.

Tony Reinhardt-Rutland

Department of Psychology, University of Ulster at Jordanstown,
Belfast BT37 OQB, United Kingdom.

Abstract

IO is a computer model designed to capture several counterintuitive features of human vision. Experiments with a novel type of Impossible Object called the Devil's Torpedo Tubes support beliefs about human vision that are embodied in IO, in particular: (1) pseudostability exists - i.e. people may notice nothing wrong with the 'objects' in anomalous pictures for a considerable time; (2) human vision does a good deal of its work in the context of local clusters of features; (3) human vision does not analyse connectivity at all thoroughly; (4) quite basic decisions in human vision are affected by transformations which change lengths and angles but leave connectivity unchanged. IO also models some quantitative features of the data, though not all.

1 Introduction

In a previous paper presented to this conference [1] we described an image interpretation system called IO which was developed to capture some distinctive features of the way people interpret line drawings of geometric objects. The system has since been refined and extended [2]. Although the domain of line drawings is limited, and in some respects artificial, it provides an opportunity to address an issue of very general significance for computational vision: how may expecations about objects' geometry be drawn into the process of interpreting visual inputs[3]?

IO is based on observations of the way people interpret line drawings. These are diverse, but the ones which are directly relevant to this paper involve the way people assign certain drawings interpretations which are geometrically inconsistent. These interpretations are popularly called Impossible Objects. One of the considerations which shaped IO was that it should attach inconsistent interpretations to the same drawings as people do, and that it should exhibit an analogue of what we have called pseudostability - that is, the state where a person interprets a drawing in a way that contains geometric inconsistencies, but is under the impression that the object is perfectly normal.

Pseudostability is a peculiarly important phenomenon because if a person is not aware of anything abnormal about the 'object' he or she is seeing, it is reasonable to assume that the processes which have led up to that point are simply the ordinary processes of vision. This is to say that pseudostable interpretations show up unexpected quirks in the normal processes of tridimensional vision. The process of discovering that an anomaly is present, and the experiences which follow that, are another matter entirely, and it is reasonable to suspect that they may reflect curious sidelines of vision with no great ecological significance. But that argument does not apply to the processes which occur up to and including the period where a person experiences pseudostability: at that stage, the drawing appears to be being processed like any other input.

Trying to model these features of vision suggests an approach to interpreting images which is natural, but has not been developed computationally before. IO's early operations form 'clusters' of junctions. Rules based on the Gestalt Laws are used to form clusters. Each cluster is then passed on to be interpreted in three dimensions. In the first instance, different clusters are assigned interpretations quite independently: but if clusters are found to be connected, then checks are carried out to establish which combinations of interpretations are mutually compatible. This outline relates to Impossible Objects in two ways. First, in certain drawings, IO interprets the fragments of an object in ways which mean that they cannot be combined into a coherent whole. This produces an analogue of the fact that people may attach inconsistent interpretations to certain pictures. Second, once local interpretations have been generated, checks are carried out to find ways of combining them into complete objects. However the checks are limited, and although they weed out most normal mismatches they allow some anomalies to go undetected. This provides an analogue of pseudostability.

This paper reports experiments designed to strengthen the empirical base on which our computational research rests. The experiments used a novel type of Impossible Object which was devised by the second author. It is based on one of the best-known

Impossible Objects, which is usually known as the Devil's Pitchfork [4]. Mitchell noticed that putting two Devil's Pitchforks back to back formed a new type of picture which he called the Devil's Torpedo Tubes. Figure 1 shows an example. This kind of figure has an intriguing property: it decomposes into two quite unconnected fragments. The right hand side of the figure forms a self contained cylinder which is not anomalous in any way, and it makes no contact at all with the lines on the left.

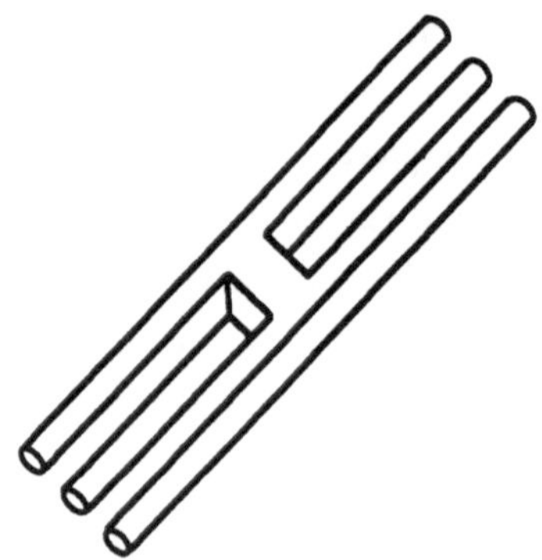

Figure 1: Devil's Torpedo Tubes type figure.

The fact that these figures decompose makes them particularly useful experimentally. It suggests a task which is straightforward from a subject's point of view, but rich in implications. It is to ask whether two picture fragments belong to the same object. This kind of task has two general attractions. One is that it feels straightforward for the subject, suggesting that responses are not overlaid by too many complex interpretive processes. That is emphatically not true of the obvious approach to probing pseudostability, which is to ask whether or not a drawing represents a consistent object. The second general attraction is that segmenting an image into separate objects is widely regarded as a basic, early step in perceptual organisation. Hence it is reasonable to argue that factors which are relevant to this task bear on fundamental steps in visual interpretation.

More specifically, the figures allow us to address several propositions about human vision which are incorporated into IO, and which it is important to back up experimentally.

The role of localised clusters. The first proposition is that human vision does a good deal of its work in the context of local clusters of features. That idea is linked to the most basic question about the Torpedo Tubes: why should they be seen as a single structure? The natural answer depends on accepting that human vision starts by breaking a picture down into fairly substantial fragments, much as IO does. In figure 1, one of the natural fragments is the cluster of junctions in the middle. If one tries to interpret that in three dimensions without bothering too much about the rest of the picture, one comes to the conclusion that it is a single structure - a sort of bar. From there it is a short step to the conclusion that the whole picture represents a single object, because everything else is connected to the bar. If that explanation is right, then the fact that the figures tend to be seen as a single structure is evidence that key decisions are carried out in the context of localised clusters of features.

Pseudostability. The second proposition is that pseudostability exists - i.e. people tend not to notice anything particularly wrong with the 'objects' in anomalous pictures unless they look carefully. IO is designed to mimic that effect - the way it subdivides the process of interpretation means it can generate descriptions which are implicitly incoherent, without detecting that the conflicts are there. That observation had a major influence on the way IO was developed, and so it is crucial to establish that the phenomenon actually does occur. The existing evidence on the point is very slight. If people are willing to report that the Devil's Torpedo Tubes form a single figure, then that would provide a reasonably clear illustration of the phenomenon.

Limited use of connectivity. The third proposition is that human vision does not analyse connectivity at all thoroughly. Traditional schemes suggest that establishing connectivity in a picture is a fundamental, early step towards working out the way a scene divides into objects. IO is based on the judgement that that is not how human observers operate. Once again, the Devil's Torpedo Tubes provide an opportunity to support that view. If humans did analyse connectivity at all thoroughly, then they could hardly miss the fact that the figure falls into two parts.

Extensive use of picture geometry. This proposition is almost a converse of the last one. In traditional schemes, basic decision-making is relatively unaffected by transformations which change lengths and angles but leave connectivity unchanged. IO reflects our judgement that human vision takes quite a different approach: it is very much affected by exact lengths and angles. Manipulating lengths and angles in the Devil's Torpedo Tubes offer a way of supporting that case. The most obvious approach involves a manipulation which is well known, but again not extensively documented. If the distance between the incompatible parts of Impossible Figures is reduced, then the anomaly can become immediately obvious, or the impression of tridimensionality can disintegrate completely [5]. The Devil's Torpedo Tubes provide an opportunity to document that kind of observation, and as will become apparent, to describe novel phenomena which are related to it.

The experiments which we report are aimed at consolidating the evidence for these propositions. They also have another function: they provide quantitative data against which models, including IO, can be matched.

A number of considerations are common to all or most of the the experiments which we report, and it is useful to set them out in advance.

1) The basic experimental technique was to place two dots on the picture, and to ask observers whether both were on the same object. This gives a way of asking whether picture fragments belong to the same object which minimises the chance of irrelevant variation in subjects' strategies. (That is a major problem with some alternative techniques, such as asking how many objects are present.) The exact position of the dots is illustrated on the stimuli in figure 2. Overall drawing size was controlled so that the dots were the same distance apart in all the test figures.

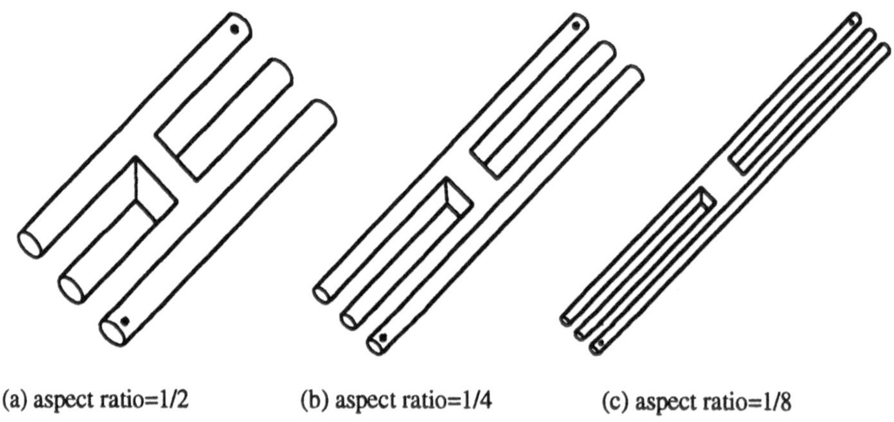

(a) aspect ratio=1/2 (b) aspect ratio=1/4 (c) aspect ratio=1/8

Figure 2: Torpedo Tubes stimuli with varying aspect ratios

2) Each subject saw only one Torpedo Tubes figure. This makes testing cumbersome, but it is necessary. Once people have tuned in to the idea that they may see anomalous objects, it is obviously unsafe to assume that their responses reflect the kind of processing that they would bring to any ordinary picture: and that is the kind of processing that we are interested in studying.

3) Three of the experiments manipulated what we call the aspect ratio of the Torpedo Tubes - that is, the ratio of the width across the tubes to the length along them (so an aspect ratio of 1 means the overall shape of the figure was roughly square, and a low aspect ratio means the figure was long and thin). The overall size of the figures was controlled by scaling them so that they all fitted into the same circle. After photoreduction, its diameter was 76.2 mm. Figure 2 illustrates the effect of varying aspect ratio.

4) Subjects were always given a sequence of four cards to get them used to the task before they saw a Torpedo Tubes figure. We call these induction cards. The induction cards all showed perfectly straightforward objects: none of them was anomalous in any way. Figure 3 shows examples.

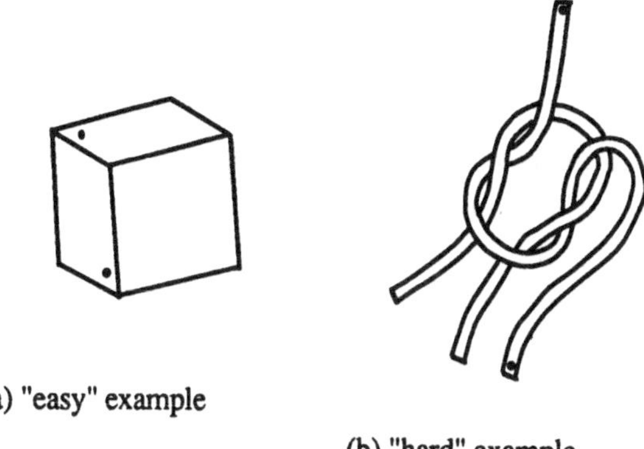

(a) "easy" example

(b) "hard" example

Figure 3: Induction cards.

5) Responses were timed. It is important to emphasise that except in the last experiment, subjects were explicitly told to concentrate on getting the decision right, not on reponding quickly. This is because we were concerned with showing that pseudostability is a reasonably enduring phenomenon, not a fleeting misjudgement caused by hurried inspection. It is not appropriate to call that sort of interval reaction time, and so we refer to it as inspection time.

2 Experiment 1

This experiment has two types of aim. One is to demonstrate phenomena which are strongly expected, but nevertheless worth confirming. The second is to validate our basic experimental technique. The two are linked: if the technique produces results which fit relatively specific expectations, then that strengthens the view that it measures what it is expected to.

The technique rests on the premise that if observers say both dots are on the same object, then they have perceived the picture as a single Impossible Object which is segmented in an anomalous way. However, subjects could conceivably make the 'same object' response for other reasons (e.g. because the dots were on the same card, or because they regarded the picture as meaningless and felt one response was as good as another).

A well known phenomenon suggests a check on the technique. Compressing drawings which suggest Impossible Objects reduces the likelihood of people accepting pseudostable interpretations; in fact, it tends to weaken the impression that the drawings represent objects at all [5]. This effect is important in its own right, as an argument for believing that human vision is very much affected by exact lengths and angles. However there are very few formal demonstrations of the effect, and it is worth confirming. The effect also offers a way of validating the two dots technique. Subjects might say that dots were on the same object for inappropriate reasons, but it is not likely that those reasons will recede as the figure is compressed - in fact, the inappropriate reasons suggested in the previous paragraph are more likely to apply in compressed drawings. Hence if 'same object' responses vary with compression in the way that is predicted, it reinforces the belief that they reflect pseudostable interpretations.

In the Torpedo Tubes figures, compression was varied by manipulating 'aspect ratio'. It was varied from 1/1 to 1/10.

2.1 Method

2.1.1 Design The independent variable was Aspect Ratio at six levels of breadth over length (1/1, 1/2, 1/4, 1/6, 1/8, 1/10), while the dependent variables were response type ('different object' v. 'same object') and inspection time (abbreviated as IT). Eight observers per cell were used, giving a total of 48 subjects. They were first-year undergraduate volunteers in the School of Psychology at Queen's University, Belfast.
2.1.2 Stimuli The test stimuli used were Torpedo Tube figures with the aspect ratios described above. Besides the test stimuli there were induction cards of the kind illustrated in figure 3(a). Each of these carried a picture which represented a straightforward 3-D solid; and each carried two red marks, placed so that making a correct response was relatively trivial.

The Electronic Developments 3-field tachistoscope which was used throughout this series of studies had a virtual distance of 508 mm., and so the maximum angular subtense of any test stimulus was always about 8.5°. Each drawing was in black outline with the two marks in red.
2.1.3 Procedure The tachistoscope was linked to a suite of timers and recorders so that a thumbswitch allowed the experimenter to replace the fixation card with a stimulus card, and a second thumbswitch allowed the subject to replace that stimulus with the fixation card. The interval between the two presses was recorded. Viewing was binocular, with the subject viewing first (and always in the same order) the three introductory cards, and then the stimulus card.
After being seated at the apparatus and set at ease, each subject was told (1) that each new drawing would appear when the experimenter pressed his thumbswitch; (2) that the task was to judge whether or not the two red spots were on the same object; (3) that inspection

time was unlimited; (4) that, having reached certainty, the subject should (a) press the second thumbswitch, and only then (b) voice the judgement, either "Yes, both marks are on the same object", or, "No, they are not".

Many Psychology students expect to respond quickly to tachistoscopic stimuli, and so care was taken to emphasise, three times during the instructions, that inspection time was free.

The four drawings were then shown, and all the judgements and inspection times noted.

2.2 Results

Results are summarised in Table 1. Response times are in seconds.

Table 1: responses in experiment 1.

	ASPECT RATIO					
	1/1	1/2	1/4	1/6	1/8	1/10
number of 'different object' responses (max=8)	8	5	1	2	2	3
IT for 'different object' responses	1.7	5.1	5.3	5.7	4.1	7.5
IT for 'same object' responses	-	1.1	1.9	2.0	3.6	3.6

The most straightforward statistical summary of the data is provided by regressions which relate responses to aspect ratio. Aspect ratio predicts response type (taking 'same object' =1, 'different object' =0). The response timings are well predicted by a model which takes aspect ratio and response type as variables: the contribution of aspect ratio is significant with $p=0.0004$, and the contribution of response type with $p=0.0003$ (i.e. 'different object' responses are significantly slower). Prediction is less good with other independent variables (notably number in the sequence, from 1-6, and the inverse or higher powers of aspect ratio); and interaction terms do not contribute.

2.3 Discussion

The results relate to two kinds of issue, procedural and substantial.

The procedural issue is whether 'same object' responses are actually indicators of pseudostability. The results reinforce the argument in two ways. First, what we know of pseudostability suggests it should fall off at low aspect ratios. 'Same object' responses are do that, i.e. they behave as we would expect if they are linked to pseudostability. Second, the other reasonably natural way of explaining that relationship conflicts with a different aspect of the data. One might suggest that people got confused with the long thin versions. However that suggests that behaviour at low aspect ratios should be random, and what we find is consensus that the points are on the same object. Those arguments reinforce the working hypothesis that 'same' responses are an indicator of a pseudostable interpretation.

If that is granted, then three substantial points can be made about beliefs associated with IO.

1) The results reinforce the belief that pseudostability does occur in human vision. Reports that the dots are both on the same object are very prevalent.

2) The results reinforce the belief that human vision uses information about connectivity in the picture in a partial and selective way - otherwise 'same object' responses would not occur.

3) The results reinforce the belief that quite basic decisions in human vision - such as decisions about whether or not points belong to the same object - are affected by transformations which change lengths and angles but not connectivity.

On a more detailed level, although there is a relationship between aspect ratio and response type and timing, inspection suggests that the relationship may not be simple. It looks not unlikely that the data follow step functions rather than smooth ones, with a break after ratio 1/2 in response type, after ratio 1/1 in 'different object' times, and (less clearly) after 1/6 in 'same object' times. The suggestion of a step function is significant because that is what models related to IO would suggest: one would expect to find steps corresponding to regions where quantitative changes in proportion result in a qualitative shifts between different ways of subdividing the figure. Less agreeably, there is also a hint that five seconds or so marks a significant divide in inspection times: subgroups which accept the anomalous interpretation look on average for less than five seconds, subgroups which reject it (except in the extreme 1/1 ratio) look on average for about five seconds. This suggests that rejection could simply be a matter of spending a critical period of time looking. Both of these points are taken up below.

Looking at the stimuli and talking to subjects gave a strong impression of the way 'different object' judgements were confirmed consciously. The key to the response was recognising that inlets at opposite ends of the figure met to form an unbroken region which divided the figure into two separate networks of lines. The important point is that secondary, consciously directed processing was needed to establish these apparently basic points about connectivity. That illustrates the kind of observation which suggested that basic processes in IO should not use much information about connectivity.

3 Experiment 2

The procedure of using induction cards makes it possible to recognise a variable which is clearly likely to affect inspection time: that is, the set that the subject brings to the task. It would be wrong to build a model round measures based on a single set of initial expectations which we have no reason to think have any special status. This experiment replicated the first, but with induction cards which encouraged more sustained inspection: an example is shown in figure 3(b). Hence it tests whether pseudostability persists even when people are strongly directed to scrutinise the stimuli carefully. Since it can be expected to generate more 'different object' responses, it should also provide more data on the timing of these responses (there were only eight of them between the four lowest aspect ratios in the previous experiment).

3.1 Method

3.1.1 Design The design replicated that of Experiment 1, using a further draft of 48 undergraduate volunteers from the same source as before.

3.1.2 Stimuli Each of the three new induction cards consisted of elements that could be described as 'worm-like' or 'ribbon-like'. These were arranged in: (1) a reef knot; (2) a regular sinusoidal double helix; (3) a regular sinusoidal triple plait. Each drawing had a maximum angular subtense of c. 8.5° at the virtual distance given by the tachistoscope; in each, the distance between the parallel sides of the elements was c. 2.5 mm.; and in each the red marks were placed at the ends of the elements in such a way that for (1) and (2) the solution was 'Same', and for (3), 'Different'. Solution necessitated careful tracking along the designated element.

3.1.3 Procedure The procedure replicated that of Experiment 1.

3.2 Results

The results are summarised in Table 2. It is clear that the new and more demanding induction cards reduced the number of 'same object' responses relative to the previous experiment: the number of responses in the category fell by half. However analysis on the same pattern as before shows that response type still depends on aspect ratio: the regression is significant with $p=0.011$. This confirms that pseudostable impressions are robust enough to produce inappropriate segmentations even when people are deliberately induced to look carefully.

Table 2: Results of Experiment 2.

	ASPECT RATIO					
	1/1	1/2	1/4	1/6	1/8	1/10
number of 'different object' responses (max=8)	8	8	4	5	5	5
IT for 'different object' responses	3.8	6.2	11.3	15.2	6.6	4.6
IT for 'same object' responses	-	-	3.0	2.5	6.7	8.0

It is also clear that no simple threshold of time spent looking determines whether pseudostability survives. Contrary to the prediction from that model, this experiment gives higher inspection times than the previous one in every cell. The shape of the data supports the model even less. At two aspect ratios the mean time for 'same object' responses is actually longer than the mean time for 'different object' responses, and the data clearly do not support the prediction that 'different object' response times should maintain a constant level (which would *ex hypothesi* be slightly above the threshold) at ratios below 1/1.

Impressionistically, two aspects of the data strengthen the case for a model based on discrete states rather than continuous relationships. First, despite considerable shifts in many respects, the data from this experiment still suggest boundaries, and they generally fall in much the same places as they did the last experiment. Second, the inspection times for 'different object' responses show a markedly non-monotonic shape.

Regressions were used to examine the statistical basis of those impressions, using a series of independent variables which have the value 1 for aspect ratios up to and including the one used in a particular card, and the value 0 above it. For brevity a variable whose value is 1 up to and including a ratio r, and zero thereafter, can be called $V \leq r$. Two phases of analysis were undertaken.

First a stepwise regression based on these binary functions was applied to the data summarised in each individual row of the tables above. This confirmed that the same break points recur throughout the data. In both experiments type of response was predicted by the same single variable, $V \leq 1/4$. In experiment 2 'same object' inspection times were predicted by the variable $V \leq 1/8$. No model met the original criteria for significance with the corresponding data in experiment 1, but when criteria were relaxed the variable $V \leq 1/8$ again emerged as a predictor. The same two variables, $V \leq 1/8$ and $V \leq 1/4$, contributed to predicting 'different object' inspection times in experiment 2. The corresponding data in experiment 1 slightly different: it was predicted by the variable $V \leq 1/2$.

Predictions based on the relevant binary functions were compared to predictions based on the continuous variable aspect ratio. The analyses covered both experiments, and an additional variable which discriminated between the two types of induction card was also used. All three dependent variables, response type, inspection time for 'same object' responses, and inspection time for 'different object' responses, showed similar patterns. The best predictor was always a model based on the directly relevant binary function or functions: however there was no great difference between models based on all three broadly relevant binary functions and those based on aspect ratio alone. Type of induction card made a significant contribution to predicting all the variables except 'different object' inspection times, where its contribution came near to significance.

3.3 Discussion

The different induction cards did reduce the number of 'same object' responses substantially. However analysis on the same pattern as before shows that response type still depends on aspect ratio. This confirms that pseudostable impressions are robust enough to produce inappropriate segmentations even when people are induced to look carefully.

The data also strengthen the case for a model based on discrete states rather than continuous relationships. For all three dependent variables (response type, inspection time for 'same object' responses, and inspection time for 'different object' responses), the best predictor was a model based on the directly relevant binary function or functions. This does not rule out models in which pseudostability is continuously related to aspect ratio, but it indicates that there is a real case for models which invoke qualitatively different representations - as IO does. The fact that breaks generally fall at the same points in the sequence is also of interest in its own right. It suggests that despite the changes produced by manipulating subjects' cognitive state, there are underlying matters of organisation which depend largely on the input. Again, this supports a key feature of IO, the distinction which it embodies between early, automatic processes and later, optional ones.

Comparing the evidence to IO suggests some natural hypotheses about the nature of the observed discontinuities. The first break (between aspect ratios 1/2 and 1/4) should correspond to the shift from an organisation which groups anomalous elements together to one which breaks the figure into three fragments, each of which can be interpreted consistently in isolation. Sightly lower aspect ratios (1/4 and 1/6) give fast 'same object' responses and 'different object' responses which are very slow: whereas still lower aspect ratios give intermediate inspection times for both. The natural explanation rests on the observation made after experiment 1, that recognising the problem depends on a distinctive strategy in which the picture is treated almost as a formal pattern, bypassing the misleading impressions which arise from considering it as an image of an object. One can then envisage that with the intermediate aspect ratios, there is a compelling and apparently coherent tridimensional interpretation which leads to fast 'same object' responses if it is accepted; and even if the tridimensional interpretation is not accepted, it keeps obtruding on the attempt to treat the object formally. At lower aspect ratios, the immediate impression of coherence is markedly weaker and does not disrupt the formal approach.

The break between aspect ratios 1/2 and 1/4 is clearly the kind of effect that IO should model. Although it does not model the contrast between a compelling sense of coherence and a weaker one, it incorporates machinery which could be applied to the problem in that the basic representation of clusters could be elaborated to a greater or lesser extent, with more or less interconnection between clusters.

4 Experiment 3

There is an obvious way of explaining why pseudostability depends on the manipulations of the figure which have been considered so far: they affect the distance between conflicting fragments (at least relative to other key distances). However the logic behind IO makes a less natural prediction: it suggests that other manipulations which have no such simple interpretation will affect the way features are grouped. Confirming that prediction would provide considerably stronger support for the ideas embodied in IO.

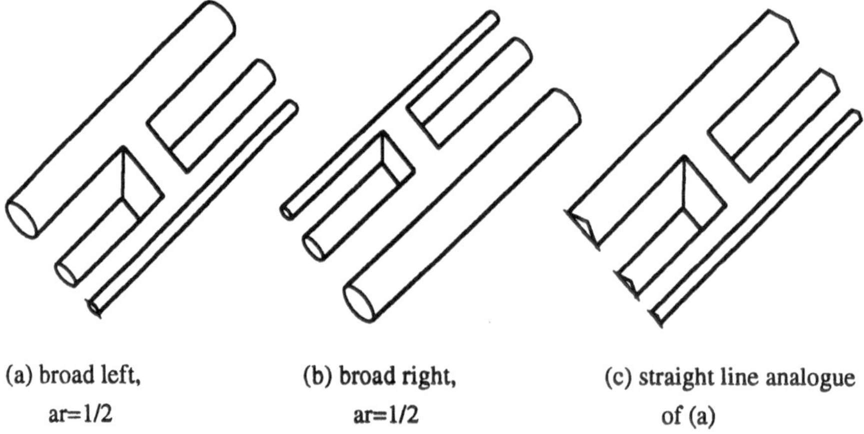

(a) broad left, ar=1/2

(b) broad right, ar=1/2

(c) straight line analogue of (a)

Figure 4: Devil's Torpedo Tubes with changing cylinder width.

This experiment examined a manipulation of that kind. Instead of using a constant spacing between the long parallels which run through the figure from bottom left to top right, we changed the spacings gradually. Two variants were considered. In the 'broad left' version, the spacings were broadest at the left of the figure and narrowed progressively as one moved right: and in the 'broad right' version, the spacings were broadest at the right of the figure and narrowed progressively towards the left. Figure 4 shows examples.

4.1 Method

4.1.1 Stimuli By this stage, it did not seem that further work with stimuli at the 1/10 and 1/1 aspect ratios would be worthwhile given the strong requirement for naive subjects. Accordingly, the 1/8, 1/6, 1/4 and 1/2 stimuli were chosen to be modified into the new test stimuli. Clarity under photoreduction dictated the 6:5:4:3:2 arithmetical series as the most extreme that could be used to control the spaces between the six parallels, this order constituting the 'broad left' series and its converse the 'broad right'. All other characteristics of the test stimuli were as in Experiment 1.

4.1.2 Design Aspect ratio was a between-subject variable with four levels (1/8, 1/6, 1/4 and 1/2); hand, also a between variable, had two levels (broad-left and broad-right). At 20 observers per cell, the design required 160 subjects, who were all undergraduate volunteers from the School of Psychology at Queen's University, Belfast, and the Psychology Department at the University of Ulster at Jordanstown.

4.1.3 Procedure As it seemed more informative to know what happened when observers scrutinised the test stimuli than when they did not, the procedure replicated that of Experiment 2.

4.2 Results

Table 3: proportions of subjects who give 'different object' responses.

	ASPECT RATIO			
	1/8	1/6	1/4	1/2
broad left	0.55	0.45	0.9	0.9
symmetrical	0.63	0.63	0.5	1
broad right	0.5	0.75	0.85	0.9

Subjects' decisions are summarised in Table 3. The results of experiment 2 are used to complete the comparison by providing data on figures with equal spacing.

Analysis of variance confirms that there is a main effect of aspect ratio (F 3,180 = 6.3, p=0.0004). An interaction which is theoretically interesting comes close to significance (F 6,180=1.8, p=0.0998). The interaction indicates that unlike the manipulation in experiment 2, changing proportions tends to shift the point where 'different object' responses become predominant. In the broad left figure, 'different object' responses predominate below and including the 1/4 overall ratio. In the broad right figure, they predominate below and including the 1/6 ratio.

Table 4: Inspection times for 'different object' responses.

	ASPECT RATIO			
	1/8	1/6	1/4	1/2
broad left	8.0	8.2	6.9	5.2
symmetrical	6.6	15.2	11.3	6.2
broad right	4.8	7.5	4.7	4.1

Inspection times for 'different object' responses are summarised in Table 4. There are significant main effects of symmetry (F 2,126 =6.88, p=0.0014) and aspect ratio (F 3,126=5.70, p=0.0011), with the interaction far from significance. The main effect of symmetry occurs because inspection times are generally highest in the figure where consistent rejection ends at the highest aspect ratio, and lowest in the figure where consistent rejection ends at the lowest aspect ratio.

4.3 Discussion

These results are not just compatible with the broad kinds of principle that underlie IO. It is possible to show quite specific similarities between the way people perceive these figures and the output of IO's clustering techniques, though the techniques were in no sense tuned to produce this outcome. Since IO cannot handle curves, the elliptical ends were replaced with triangular ones as in Figure 4 (c), and these figures were used as inputs to the 'cluster and divide' part of IO.

With the symmetrical figures, IO makes a qualitative distinction in roughly the right place. When aspect ratio is 1/1, IO finds a single cluster. According to the argument sketched in the introduction, that should preclude pseudostability. At higher aspect ratios, the figure is divided into three - a rectangular midsection and two ends, each composed of three triangular tines. That should produce a pseudostable interpretation, which corresponds to the empirical evidence well except in the case of the 1/2 aspect ratio. Both asymmetrical figures continue to give a large cluster, which should not give rise to pseudostability, down to lower aspect ratios - again corresponding to the empirical evidence. However the actual position of the break point is wrong again: IO would not find clusters which led to pseudostability until the aspect ratio was below 1/10.

It noteworthy that here IO is right about a relationship which an intuitive approach would mispredict. One might expect that the key variable would simply be the breadth of the 'channel' which divides the right hand tine from the rest of the figure. That correctly predicts that the broad right figures will receive more different object responses than the broad left figures, but it also predicts that the symmetrical figure should be intermediate between the broad left and broad right variants. That clearly does not happen. Instead as IO predicts, both asymmetrical variants shift to a preponderance of 'different object' responses at lower aspect ratios than the symmetrical variant does.

This level of match has to be considered encouraging for the general approach behind IO. Equally significant is the point that the data present model-builders with targets which are non-obvious but reasonably clear.

5 Experiment 4

The first experiments underlined the fact that pseudostability lasts. However there is a critical question at the other end of the time spectrum. IO depends heavily on the assumption that pseudostability arises because normal interpretation procedures do not uncover anomalies in Impossible Objects. There is another possibility, though: that normal analysis procedures find anomalies, decide that they are insoluble, and retreat from them. If that is so, then Impossible Objects are not a particularly interesting experimental tool.

This experiment tested a key prediction from the hypothesis that anomalies are picked up early and automatically, but papered over. If so, then speeded responses should be slower with an Impossible Object than they are with a matched straightforward picture. Aspect ratios of 1/10, 1/6 and 1/2 were chosen to sample what appear to be the key parts the range. At each aspect ratio, a symmetrical Devil's Torpedo-Tube from Experiment I was paired with a closely matched drawing which represented a straightforward H-shaped object.

5.1 Method

5.1.1 Stimuli Aspect ratios of 1/10, 1/6 and 1/2 were chosen to sample what appear to be the key parts the range. At each aspect ratio, a symmetrical Devil's Torpedo-Tube from Experiment I was paired with a drawing which was as closely equivalent as possible, but which fully bore the interpretation of being a stable 3-D object. This was done by re-connecting the ends of each drawing so that it showed a symmetrical H-shaped structure composed of rectangular beams. Each of the new 'H' stimuli was prepared to the same specifications as the test stimuli from Experiment 1.

5.1.2 Design A between-subject design was used, with three levels of Aspect Ratio (1/10, 1/6 and 1/2) and two levels of Pseudostability, (stable [H], and unstable [Torpedo-Tubes]). There were 12 observers per cell, making 72 subjects in all. They were naive first-year subjects from the School of Psychology at Queen's University, Belfast.

5.1.3 Procedure This replicated that of the previous experiments except that the wording of the intructions was remodelled in one respect: each of the three assurances that inspection time was free was replaced by an injunction to make the judgement as quickly as possible.

5.2 Results

Table 5: Frequency and inspection time for 'same object' responses to Impossible (i.e. Torpedo Tubes) and Possible (i.e. H shaped) type objects in Experiment 4.

	ASPECT RATIO					
	1/2		1/6		1/10	
'same object' responses	number	IT	number	IT	number	IT
Impossible	7	0.89	11	0.92	9	0.95
Possible	12	0.66	11	0.99	10	1.18

Table 5 summarises subjects' responses. Unsurprisingly, there is a trend for the possible object to receive more 'same object' responses which almost reaches significance (F 1, 66=3.8, p=0.055). There is also a trace of the familiar finding that low aspect ratios lead to fewer 'same object' responses, reflected in an interaction which is somewhat further from significance (F 2, 66=2.22, p= 0.12). Applying a regression to the timing data as before shows that responses slow significantly as aspect ratio increases (F 1,58=5.0, p=0.030). However object type does not affect timing, and indeed apart from the 1/2 ratio 'same object' responses to the impossible object are marginally faster than responses to the possible one.

5.3 Discussion

Although the data from this study are not conclusive, they are as our view of pseudostability predicts. The experiment may be too insensitive to detect the slowdown caused by detecting anomaly and deciding to ignore it, but that argument is weakened by the fact that it is sensitive enough to detect differences associated with aspect ratio. So while it remains possible in principle to argue that anomalies are picked up early and automatically, our approach certainly fits the available evidence more naturally.

6 Overall discussion

Our empirical work reinforces the view that early processes in vision build representations which are like IO's in some interesting senses. The general characteristics of Torpedo Tubes type figures indicate that in human vision as in IO, a good deal of work is done at the level of clusters, and limited use is made of connectivity. Finer grained information from the experiments confirms that continuous changes in figures' proportions lead people to adopt qualitatively different organisations, as happens in IO's clustering; that there are broad correspondences between the way proportion affects people's choice of organisation and the way it affects IO's; and that like IO, people simply do not detect certain kinds of contradiction in the representations which they build. This is useful support for the general approach that IO embodies.

Along with this support come a number of details which do not fit IO. This is not surprising, since we make no claim that IO is a finished model. It is also encouraging in an important sense: it confirms that this is an area where empirical evidence can be collected and can provide interesting constraints on model building.

Acknowledgement: this research was funded by grant G/RE 88097 from the U.K. Science & Engineering Council.

References

[1] Cowie, R., Hamill, T., Morrow, P. and Perrott, R. IO: towards an alternative image of human vision. In A. Smeaton & G. McDermott (eds.) *Artificial Intelligence and Cognitive Science 89* Springer Verlag British Computer Society Workshop series, London, 1990.

[2] Cowie, R., Hamill, T., Morrow, P., Perrott, R., Mitchell, R. and Reinhardt-Rutland, A.. *3-D interpretation of contours in a single image.* Final report to SERC on grant G/RE 88097, 1989.

[3] Cowie, R. and Clements, D. The logical basis of visual perception: Computation and empirical evidence. *Irish Journal of Psychology* **10**: 232-246, 1989.

[4] Schuster, D.H. A new ambiguous figure: the three-stick clevis. *American Journal of Psychology* **77**: 673, 1964.

[5] Hochberg, J. In the mind's eye. In R.N.Haber (ed) *Contemporary Theory and Research in Visual Perception.* Holt, Rinehart & Winston, New York, 1968.

Deviant epistemics

John G. Harper
Computer Science Department
St Patricks College, Maynooth
IRELAND

April 12, 1991

Abstract

This paper outlines an extension to the Q modal systems based on epistemic interpretations of weak and strong necessity operators. The resulting system, Q_{EP}, is a complex three-valued epistemic logic. The introduction of truth-value gaps for certain formulae, arguably, accords with native intuitions that in particular contexts certain beliefs may lack sufficient evidence for or against their obtaining. Such beliefs are deemed irrefutable in the epistemic extension to the Q systems. Some comments on the theorem proving problems likley to be encountered are outlined in the penultimate section. Parallels between this work and some recent work on liklihood are stated in the conclusion.

1 Introduction

Formalized theories of knowledge and action have important consequences for the design of autonomous intelligent systems. Common difficulties with most formal theories of beliefs are firstly, the inability to avoid the attribution of logical omniscience to the believing agent (if an agent can make any inference then they should *logically* make inferences compatible with that inference. As experience shows, however, this view of logical consequence is not reconcilable with the ordinary exercise of intelligent behaviour), and secondly, the presupposition that propositions are either true or false. Application of the principle of bivalence rules out treating suppositions or hypothetical statements as "unproven". The approach outlined here rejects bivalence and introduces the notion of semantic supposition for statements deemed unproven. On the other hand, logical omniscience is only partly solved. The formal vehicle is an epistemic interpretation of a modal logic wherein the modal operator M for weak necessity is interpreted as an epistemic operator B_{Mx} for x *believes that* [1].

[1]The epistemic analogues of Q modal operators used here are B_M and B_S. An exact description of each is provided later while distinguishing the modal Q systems from the epistemic system Q_{EP}.

Thus the epistemic formula $B_{Mx}(\rho)$ is intended to convey that the agent x believes that the proposition ρ *is true*[2].

2 Historical background

Model theory is concerned with specifying the circumstances under which sentences of a formal theory are legitimate. If a sentence is true in one interpretation, it is deemed *satisfiable.* A sentence which is satisfiable in every interpretation of a theory Γ is termed *valid.* These principles obtain for any application of model theory, be it in standard classical logic or non-standard modal logic. It is the latter which is of interest, and the approach adopted here is motivated by Kripke [4]. In a standard Kripke structure ***true in all possible worlds*** and ***not false in all possible worlds*** are logically equivalent. This relationship obtains on the assumption of the validity of the principle of bivalence, viz. a formula is either true or false. Early work by Prior, [5,6], questioned this assumption by introducing a series of logics which permitted truth-value gaps known as the Q systems. For all intents and purposes, these were strict three-valued systems, i.e. based on *strict* implication. Ruzsa [8,9] provided the Q systems with a set theoretic semantics. Work by Kripke [4] and more recently Turner [11], though with very different intent, have suggested that in terms of oiling the wheels of a theory of truth and belief bivalent modal systems are weaker than their three valued counterparts. More significantly, Turner [11] proves that any formal theory of truth and modality meeting his sufficiency criteria must eschew naive interpretations of bivalence[3]. There are good grounds for defending the thesis that polyvalent rather than bivalent logics offer a better purchase on truth and modality as portrayed in ordinary discourse. The core of this argument is that commonly discourse participants must admit "unproven" statements into their exchanges in two senses. In the first case a statement may be scientifically or formally unproven. While in the second case a statement may be unproven in that the knowledge possessed by two participants is insufficient to prove the statement either true or false. For example, the problem of *future contingents* such as "(I believe that) there will be a sea battle tomorrow" was well known in antiquity. Of most interest here are statements with a similar status, termed ***suppositions.*** In the system sketched below these are defined as semantically irrefutable relative to a given model – a given set of possible worlds. A formula is irrefutable if its negation can be proved unsatisfiable within any model accessible to it (and by implication within any compatible one). However, the unsatisfiability of $\neg\rho$ does not imply ρ is "always true", rather ρ is *irrefutable.* Thus, an inference from the unsatisfiability $B_{Mx}(\neg p)$ to $B_{Mx}(p)$ will be invalid depending on context[4]. Likewise, and hardly novel, the modal principle $\neg\Diamond(\rho) \rightarrow \Diamond(\neg\rho)$ has no epistemic analogue in Q, i.e. $\not\models \neg B(\rho) \rightarrow B(\neg\rho)$.

[2]Not to be confused with $B(\rho) \longrightarrow \rho$ which generates a degraded $S5$-type system, i.e. ρ is true if x believes ρ; but $\rho, \vdash_Q B(\rho) \equiv_{def} K(\rho)$ states that if ρ is demonstrably true and believed to be so then the epistemic agent ***knows*** ρ is true.

[3]This implies if stated as $true(A) \vee fasle(A)$ that bivalence must be given up. Turners's own axioms for truth **DIS** $\neg(T(t) \wedge F(t))$ and **FT** $F(A) \Leftrightarrow T(\neg A)$ still reflect the sensible notion that truth and falsity are disjoint.

[4]Technically, context is defined by either a ***frame*** $\langle W, R\rangle$ or a ***designated frame*** $\langle W, R, w\rangle$ where R is a class of accessibility relations and w is a designated world. The interpretation of a formula is therefore conditioned by the set of accessibility relations operating between worlds.

Vars. $p\backslash q$	$p \wedge q$ T I F	$p \vee q$ T I F	$p \rightarrow q$ T I F	$p \Leftrightarrow q$ T I F
T	T I F	T I T	T I F	T I F
I	I I I	I I I	I I I	I I I
F	F I F	T I F	T I T	F I T

p	$\neg p$
T	F
I	I
F	T

Figure 1: Kleene 3-valued *weak* truth tables

The exclusion of this rule is important on the grounds of plausibility, otherwise an agent would be committed to believing ρ or $\neg\rho$.

In what follows here, a first-order epistemic extension to the Q systems is developed, termed Q_{EP}. In keeping with the theme of Ruzsa's and Turner's work the semantics is based on Kleene's 3-valued system in Figure 1. The difference, however, is that Q_{EP} the epistemic interpretation of Q emphaises the *weak* interpretation of the logical connectives rather than Kleene's *strong* interpretation. Only a characterisation of the weak system is possible here. If any of the arguments to a weak connective is unstatable then the resulting formula is also unstatable. Quite literally this means that if a formula contains a truth-functionally indeterminate component, the whole formula is indeterminate. The actions of the weak connectives are therfore similar to Bochvar's "internal" connectives. Despite the sometimes wearying hair-splitting within 3-valued logics, the computational worth of these approaches has been underlined by Turner[5] according to whom, a robot using such systems as a basis for reasoning would be "well-behaved" and not likely to support unwarranted conclusions. Not all of the many semantic and computational problems are addressed below, but some flavour is given of the possibilities of using a 3-valued epistemic logic.

3 Description of the Q systems

The language of the Q modal systems contains the symbols for negation, implication and the modal operators M and L (possibility and strong necessity, respectively). An interpretation for a set of formulae P is formed from the triple $\langle W, R, V\rangle$ where W is a nonempty set of possible worlds, R is an *accessibility* relation on W and V is a two-place valuation function from $P \times W$ into the set $0, 1, 2$. By convention '0' and '1' represent F, 'false', and T, 'true'. The third value '2', I, represents a truth-value gap,. Intuitively, this value affirms the unstatability of a sentence's truth or falsity. In the epistemic extension this implies that the modal expression $M\rho$ will be interpreted as $B_M(\rho)$ which is intuitively rendered as "it is believed that ρ". The usual logical conventions defining the formal syntax of a language are assumed.

The divergence between the Q systems and classical modal logic is quite marked. For instance, in most modal systems $\neg M \neg\rho$ and $L\rho$ are equivalent. In the Q systems not so. The inference from the impossibility of $\neg\rho$ to the strong necessity of ρ is unsound. Hence, in $L(\rho \rightarrow \rho)$ the *statability* of $\rho \rightarrow \rho$ is being expressed. Only the possibility of its being false is excluded. It is not a tautology. Classically, a

[5][7] and Turner[10,11] investigate the strong Kleene connectives and their similarity to Bochvar's connectives. Both these systems are *monotone*. A mapping between Kleene's weak system and Bochvar's preserves monotonicity.

formula is unstatable if it lacks a truth-value at a possible world w. By implication if a variable is uninstantiated at a world, every formula containing that variable is unstatable at the same world (and straightforwardly false in most cases). But of more particular interest is the case where a predicate fails to have an extension at a possible world due to the structure of the predicate itself. For instance, the attribution of a particular property to a domain may simply lead to an ambiguous proposition, e.g. "six is the brownest number". Using Ruzsa's terminology such predicates are termed *degenerate*. Various treatments have been proposed over the years for statements with opaque extension. Suffice it to say that it is still a considerable problem, perhaps more seriously for possible world semantics as will be hinted at below. At the moment, careful distinctions between a predicate's opaque and non-opaque contexts have been avoided. However, this would not suffice for a full epistemic extension of Q allowing quantification over singular terms.

3.1 Logical equivalence in the Q systems

The original sin of possible worlds arises as follows: sentences which are logically equivalent express the same proposition, say **TRUTH**. Under any equivalent substitution in an extensional framework **TRUTH** therefore remains invariable

$$\forall w \in W, if val(m_i) = val(m_j) \Longrightarrow \forall w, val(\rho(m_i))_w = val(\rho(m_j/m_i)_w)$$

where $val(X)$ is the semantic value of X. This result poses particularly difficult problems for logics dealing with propositional attitudes. In the ordinary course of events, propositions are the objects of *believing*, *knowing*, *estimating* and so forth, and are therefore distinct from each other in the sense that one may be believed while another may not. But if a set of propositions is logically equivalent and thus true in the same sets of possible worlds, then believing one proposition logically entails a commitment to **TRUTH**, the set of logical equivalents.

The nub of the problem is that the *facts of entailment* do not in common experience accord with the *logic of entailment*. Speakers are generally ignorant of the variety of logical relationships implied by their utterances. Logics of belief must reflect this fact. The problem has been aptly characterised as one of *logical omniscience*, and it is doubtful whether any formal theory basing its semantics solely on possible worlds can entirely avoid it.

How does this affect the development of an epistemic extension to the Q modal systems? In the first place, there are inherent limitations in any theory using possible worlds to found a logic of attitudes. The worst effects can be mitigated somewhat by imposing a restrictive schema for substitution among logical equivalents, and this is followed in Q systems. Secondly, the explicitly extensional nature of the Q systems, in this instance anyway, entails no comment on what a proper natural language semantics should provide. Some may consider this a serious deficiency, arguing that a theory of attitudes is intrinsic to a semantic theory. Finally, not just *any* three-valued logic will fit the bill. The compatability of Q_{EP} with the Kleene connectives renders it at least computationally plausible. Moreover, in the light of empirical evidence imposing a restraint on the application of bivalence is suggestive of some heuristic adequacy.

3.2 Satisfaction and statability in Q_{EP}

Much of what follows hinges upon the interpretation of the the notion "*x believes that p*". Agent x must hold as inconceiveable that p is *statable by x* and yet supposed as false by x. This accords well with intuition, as it is logically contradictory for an agent to simultaeneously aver to the truth or irrefutability of a proposition and yet suppose the proposition to be false. An agent's belief is *completely warranted*, using crude modal analogues for the moment, as follows:

$$B_{Lx}(p)_{def} \equiv \neg(B_{Sx}(p) \rightarrow B_{Mx}(\neg p))$$

Using the Q notions for necessity L, statability S and possibility M, this states that a belief is warranted if and only if an agent believes that the belief is statable and not false.

Replacing $\neg B_{Mx}\neg$ with B_{Suppx}, a strong version of an $S5$-type axiom for *x knowing p* is an immediate consequence by *epistemic necessitation:*

$$p, \vdash B_{Lx}(p) \equiv p, \vdash (B_{Sx}(p) \wedge B_{Suppx}(p)) \equiv_{def} K_x(p)$$

An agent therefore knows that a propositon is true if (a) the proposition is actually true in a world accessible to the agent, (b) the agent can state that the proposition is true (and at least temporarily irrefutable), and (c) will not admit the proposition as a false supposition. In all cases B_S is the operator expressing epistemic statability. To interpret sentences containing the operator a set of valuation rules must be provided wherein each symbol or symbol-complex receives a unique semantic value. An interpretation for a Q system consists of the quadruple $I = \langle W, R, U, \xi \rangle$, where W is a set of possible worlds, R is a class of accesibility relations, U is a domain of distinguished individuals, and ξ is a valuation function, e.g. the value of $\xi(p)$ at some $w \in W$ is $\xi(p)_w$. A parameter $\rho \in \Pi$ is *degenerate* at some $w \in W$ if and only if:

1. $\rho = \forall x_1, ..., x_n, ...$, and $\xi(\forall)_w = 0$[6], or
2. $\rho = e$, where e is an individual parameter, and $\xi(e) \notin \xi(\forall)_w$, or
3. ρ is a predicate symbol, and $U \in \xi(\rho)_w$.

The last definition of degeneracy addresses the paradoxes arising from the use of formulae expressing properties of totalities. If the extension of any predicate is the total set of individuals U then that expression is degenerate. Furthermore, if any part of a formula is degenerate at a particular world w, then the formula as a whole is degenerate for the same world, and thus unstatable[7]. A predicate is degenerate with respect to Q_{EP} just in cases when it is degenerate with respect to Q. This equivalence can be strengthened for cases relative to the actual world. The main

[6]In other words if some variable in the scope of the universal quantifier does not have a denotation at w then ρ is degenerate and any sentence α containing ρ is truth-functionally *unstatable*. The assignment $\xi(\forall)$ is a function from the set of possible worlds W into the power set of U. At any world $w \in W$ the value of $\xi(\forall)$ is $\xi(\forall)_w$. The denotation of an expression fails at a cetain world w_v when some subset of U fails to hold in w_v.

[7]This accords with the Frege-Russell-Quine principle that sentences with "fictional" names lack truth-value, and are semantically unstatable.

function of the treatment above is to provide some purchase on situations which use nondenoting terms. A robot confronting such utterances can suspend judgement on their truthfulness.

The rules for the connectives can be read from Figure 1. The Q_{EP} valuation rule for suppositionhood, below, is not merely a restatement of the modal principle that if a proposition is possibly true then it is a legitimate supposition. The rule proceeds by way of an elaboration of the semantic effect of the provability or otherwise of the *negation* of a proposition:

$Q_{EP} - Supp$: If $\xi(\neg\rho)_w = 2$ then $\xi(B_{Suppx}\rho)_w = 2$. Otherwise, $\xi(\neg\rho) \neq 2$ implies $\xi(\neg\rho)_w = 1 - \xi(\rho)_w$ such that $\xi(B_{Suppx}\rho)_w = 1 - \xi(\neg\rho)_w$.

Three obvious but distinct cases are covered by this rule[8] : (a) if $\neg\rho$ is unstatable then $B_{Suppx}(\rho)$ is also unstatable, (b) if ρ is true then so is $B_{Suppx}(\rho)$, and (c) if ρ is false then likewise for $B_{Suppx}(\rho)$.

The semantic rule for statability emphasises that if a proposition's truth-value is not unstatable at world w but is unstatable at v which is accessible to w then the truth-value of the statement "agent x believes ρ" is false.

$Q_{EP} - S$: If $\xi(\rho)_w \neq 2$, and for $v \in W$, both $\langle w, v\rangle \in R$ and $\xi(\rho)_v = 2$ obtain, then $\xi(B_S\rho)_w = 0$. On the other hand, if $\xi(\rho)_w = 2$, then $\xi(B_S\rho)_w = 2$. In all other cases, $\xi(B_S\rho) = 1$.

The rules above seem to capture basic intuitions speakers have about belief and judgement. For instance, in the case of suppositions if it is assumed that their truth-value is unknowable, or at least unknowable relative to the world of the speaker, then the truth-value of a belief about a suppostion is also indeterminate. The rule governing statability seems odd at first glance but it is simply a variant on the modal rule for necessity.. It places a strong restriction on attempts to mimic "paraconsistency". If a proposition is possible at world w, $\Diamond(\rho)_w$, but impossible at world v accessible from w, $\neg\Diamond(\rho)_v$ then $\xi(B_S(\rho))_w = 0$[9]. By way of illustration, if at world w it can be proved that the proposition "the moon is made of cheese" cannot be truth-valueless, but an agent with access to world v advances a contrary argument successfully then agents in w must accept that "we believe that the moon is made of cheese" is false in w. At first this may jar with intuition, afterall the agents in world w are being forced to believe something for which they have no evidence. But is this any different from experience? The emphasis is upon preserving truth across worlds rather than diluting it. As a general principle, if the truth-value of a formula is undecidable at a world v but known not to be undecidable at an immediate ancestral world w (without a positive evaluation of the truth-value), then any embedding of the formula in an epistemic context in w renders it refutable, e.g. $\xi(B_S(\rho))_w = 0$. Using the rules above a further rule for warranted belief can be derived if required.

[8]In fact a sleight of hand was performed here to simplify the presentation. The more correct statement of the rule entails appeal to the *possible* truth or falsity of ρ, i.e.

$$\ldots\xi(\neg\rho) \neq 2 \text{ implies } \xi(\neg\rho)_w = 1 - \xi(\rho)_w \text{ such that } \xi(B_{Suppx}\rho)_w = 1 - \xi(M\neg\rho)_w.$$

[9]Work is underway to move most of the Q_{EP} apparatus over into Turner's [11] framework, with statability recast as a truth predicate.

For the sake of completeness two further rules must be added for quantified expressions which "fall over", i.e. denotation failure occurs. The set U designates a set of possible denotations for variables, individuals and predicate expressions. Occasionally an expression may fail to pick out a referent at a particular world because the formula in which it occurs is truth-valueless. The following rules capture the quantificational effect of a proposition's unstatability[10] :

$Q_{EP}\forall$: If $\xi^{u/x}(\rho)_w = 2$ for all $u \in U$ then $\xi(\forall x\rho)_w = 2$. If for *any* $u \in U$, $\xi^{u/x}(\rho)_w = 0$ then $\xi(\forall x\rho)_w = 0$. Otherwise, $\xi(\forall x\rho)_w = 1$.

$Q_{EP}\exists$: If $\xi(\forall x\rho)_w = 2$, then $\xi(\exists x\rho)_w = 2$. If $\xi^{u/x}(\rho)_w = 1$ for any $u \in U$ then $\xi(\exists x\rho)_w = 1$. Otherwise, $\xi(\exists x\rho)_w = 0$

Hence, a sentence (a quantified formula with no free variables) inherits the truth-value of its base formula in the general case.

A sentence ρ is a consequence of a set of sentences Γ if and only if $\{\Gamma, \neg\rho\}$ is unsatisfiable. A proof progresses by refutation. If ρ is refutable then it must be the case that $\neg\rho$ is satisfiable. An important corrollary is that if ρ is *irrefutable* then $\neg\rho$ and $\neg B_{Suppx}(\rho)$ are unsatisfiable. Consequently, every supposition of an irrefutable sentence is itself irrefutable, i.e. if $-B_{Suppx}(\rho)$ is unsatisfiable then $B_{Suppx}(\rho)$ is irrefutable. This accords with the principle that if a proposition cannot be proved false then it is accepted as epistemically irrefutable in an argument. A further strengthening of this rule is as follows: assume the set of epistemic sentences Γ includes the sentence ρ, but $B_{Lx}(\neg\rho)$ is not a consequence[11]. The sentence $B_{Suppx}(\rho)$ is then a consequence of Γ. The formal expression is rendered

$$\frac{\Gamma, \vdash \rho \qquad\qquad \Gamma, \nvdash B_{Lx}(\neg\rho)}{\Gamma, \vdash B_{Suppx}(\rho)}$$

3.3 Irrefutability, truth and belief

So far there has been no comment on the universality of the rules of inference of classical logic. Classical systems rely on the *modus ponens* to effect inference because it is provably the case that for classical logics the rule is truth-preserving. However, in the case of Q_{EP} it is by no means obvious which semantic properties are preserved by the *modus ponens*. The rule as stated for standard systems is as follows:

$$\frac{\Gamma, \models \rho \qquad\qquad \Gamma, \models \rho \rightarrow \beta}{\Gamma, \models \beta}$$

If ρ is a consequence of Γ and $\rho \rightarrow \beta$ is also a consequence of Γ then β is a consequence of Γ. The rule is truth-preserving, for it cannot be the case the ρ is true, and $\rho \rightarrow \beta$ is true and yet the consequence β be false. Bivalence assures this is the case, but does a polyvalent system entertain the same strict notion of semantic inheritance?

[10]Assume formulae are closed under quantification. Unground variables complicate the quantificational apparatus by failure in rigid designation. Thus if $\forall$ and $\exists$ are considered as functions from W into the $Pow(U)$ then $\xi(\exists x\rho)_w$ may not be logically equivalent to $\xi(\neg\forall x\neg\rho)_w$ in all cases. Readers are referred to Ruzsa [9] for some further comment.

[11]No distinction is drawn between syntactic and semantic consequence at the moment. In a classical system the two properties "coincide" in a completeness theorem demonstrating the coextensionality of provability and truth.

In standard modal Q irrefutability is not preserved by *modus ponens* but truth is maintained, on the other hand. A tautology in a bivalent system is simply one which is true under any and every assignment of truth-values. This is *not* the case for many-valued logics. Tautologies are produced by seeding or *antidesignating* truth-values. Appropriate designations will produce tautologies. Hence, a set of axioms for a many-valued system are only tautologous under selected truth-value designations[12] [7]. By way of illustration, the classical rule of necessitation:

$$\Box(\rho \wedge \varrho) \supset \Box(\rho)$$

is semantically very weak as a derived proof rule in all the Q systems. To establish *refutability* it is necessary to prove that a sentence is not irrefutable, i.e. $\xi(\Phi) = 0$ is possible. As an illustrative case the following proof is adduced

1. Assume $\rho \neq 2$ and $\varrho = 2$
2. $\Box(\rho \wedge \varrho) = 2$ — from 1, $\Box$ and $\wedge$ rules
3. $\Box(\rho \wedge \varrho) \supset \Box(\rho) = 2$ — from 1,2 and $\supset$ rule
4. $\Box(\rho) \neq 2$ — from 4 and *modus ponens* rule

Thus, while A and $A \supset B$ are irrefutable B does not necessarily inherit that property. Significantly, the same conclusion does *not* hold for the epistemic formula

$$B_{Suppx}(\rho \wedge \varrho) \rightarrow B_{Suppx}(\rho)$$

This result has very important epistemic implications for the simple reason that certain argument schemata involving suppositions are irrefutable in principle. The semantic rules for B_{Supp} state that $B_{Supp}(\rho)$ is true when $\neg\rho$ is false, and false when $\neg\rho$ is true. An intelligent agent should, intuitively at least, accept that if a conjunction of propositions is supposed, then at least one of the conjuncts can be separately supposed. More directly, it is important in reasoning about knowledge to be able to state whether, on the basis of a prior supposition-containing argument, a belief is warranted or not. Weakening some of the stronger rules of necessitation places an upper limit on the distribution of "supposition-hood" in a compound formula.

4 Completeness and proof by tableaux

The systems under the umbrella of Q_{EP} are much "denser" than than those standardly derived from $S4$ and $S5$ logics. The treatment of unstatable and irrefutable propositions within the various accessibility relations R is the main source of complexity. Despite this, Q_{EP} is complete even up to its $S4$ and $S5$ interpretations. Proving that every maximal consistent set in Q_{EP} is satisfiable in any interpretation it defines is no ordinary matter. The present completeness proof for Q_{EP} relies for the moment on a strenuous adaption of the Henkin method. The main computational task at the moment is constructing a sound theorem prover. One problem relates to the treament of negation. Classically, if any branch in a tableau

[12]The absence of Q_{EP} axioms in this paper is deliberate

contains $[\rho, \neg\rho]$ a contradiction is obtained and that branch is closed. In a three-valued system determining branch closure conditions is not as simple, and is even less so in Q_{EP}. Moreover, in Q_{EP} it obviously makes a very important difference whether a formula has been determined as closed under *irrefutability* or closed under *contradiction.* For instance the semantic valuation of a disjunction is unstatable and equal to the that of a conjunction if either one of the disjuncts, or conjuncts, is itself unstatable. A prover based on bivalence can simply conclude that if $\neg\rho$ is unsatisfiable, then ρ is true. It would appear that negative literals in a Q-formula pose a problem and a Q prover may, on the other hand, merely conclude that ρ is *irrefutable.* At the moment it is speculated that appropriate antidesignations will be needed. Secondly, *modus ponens* in Q_{EP} does not preserve irrefutability except in the cases of a number of argument schemata involving suppositions, however, it does preserve truth. If this was not the case, the Q_{EP} systems would be inconsistent.

5 Conclusion and parallels with other work

At the moment a theorem prover for *both* propositional and predicate cases is being built which is complicated by the failure of the standard quantifier equivalences under certain conditions. A preliminary propositional prover is seeded towards generating branch components with clear refutability properties for practical reasons. This has been influenced by Fitting's work on tableaux [1].

Parallels with the work briefly documented here is that on the logic of liklihood **LL** by Halpern and Rabin [2] and Konolige's deductive belief models [3]. These latter efforts are exclusively supportive of bivalence. For reasons of formal purity, the Q apparatus will eventually move over to Turner's property theory paradigm. Firstly, possible worlds are unstable and the problem of logical omniscience cannot be avoided with consistency. Secondly, Turner's theory provides for the retention of those aspects of bivalence which are sensible, especially in a computational paradigm.

Despite the fact that intuitive accounts of native conversation seem to require some account of "truth suspension", very little current research is focused on providing the necessary formal theories. Certainly, the problematic status of the laws of the excluded middle, double negation and non-contradiction in such systems present problems for theorem provers. But are we *that* sure we have bivalence right?

6 Bibliography

1. Fitting, M. (1990). *First-order logic and automated theorem proving.* Springer-Verlag.

2. Halpern, J. Y. (1987). "A logic to reason about liklihood", *Artifical Intelligence* ,**32**, 379–405.

3. Konolige, K. (1986). *A deduction model of belief.* Morgan Kaufman, Los Altos.

4. Kripke, S. (1975). "Outline of a theory of truth", *Journal of philosophy,* **1**, 690–716.

5. Prior, A. N. (1957). *Time and modality.* Claredon Press, Oxford.

6. Prior, A. N. (1964). "Axiomatizations of the modal calculus Q", *Notre Dame J. Formal Logic,* **5**, 215–217.

7. Rescher,N. (1968). *topics in philosophical logic.* D. Reidel.

8. Ruzsa, I. (1973). "Prior-type modal logic, I-II. ", *Periodica Mathematica Hungaria,* 4, 51–69, 183-201.

9. Ruzsa, I. (1981). *Modal logic with descriptions.* Martinus Nijhoff, The Hague.

10. Turner, R. (1985). *Logics for artificial intelligence.* Ellis-Horwood.

11. Turner, R. (1990). *Truth and modality for knowledge representation*, Pitman, London.

Author Index

Published in 1990–91

AI and Cognitive Science '89, Dublin City University, Eire, 14–15 September 1989
A. F. Smeaton and G. McDermott (Eds.)

Specification and Verification of Concurrent Systems, University of Stirling, Scotland, 6–8 July 1988
C. Rattray (Ed.)

Semantics for Concurrency, Proceedings of the International BCS-FACS Workshop, Sponsored by Logic for IT (S.E.R.C.), University of Leicester, UK, 23–25 July 1990
M. Z. Kwiatkowska, M. W. Shields and R. M. Thomas (Eds.)

Functional Programming, Glasgow 1989, Proceedings of the 1989 Glasgow Workshop, Fraserburgh, Scotland, 21–23 August 1989
K. Davis and J. Hughes (Eds.)

Persistent Object Systems, Proceedings of the Third International Workshop, Newcastle, Australia, 10–13 January 1989
J. Rosenberg and D. Koch (Eds.)

Z User Workshop, Oxford, 1989, Proceedings of the Fourth Annual Z User Meeting, Oxford, 15 December 1989
J. E. Nicholls (Ed.)

Formal Methods for Trustworthy Computer Systems (FM89), Halifax, Canada, 23–27 July 1989
Dan Craigen (Editor) and Karen Summerskill (Assistant Editor)

Security and Persistence, Proceedings of the International Workshop on Computer Architecture to Support Security and Persistence of Information, Bremen, West Germany, 8–11 May 1990
John Rosenberg and J. Leslie Keedy (Eds.)

Women Into Computing: Selected Papers 1988–1990
Gillian Lovegrove and Barbara Segal (Eds.)

3rd Refinement Workshop (organised by BCS-FACS, and sponsored by IBM UK Laboratories, Hursley Park and the Programming Research Group, University of Oxford), Hursley Park, 9–11 January 1990
Carroll Morgan and J. C. P. Woodcock (Eds.)

Designing Correct Circuits, Workshop jointly organised by the Universities of Oxford and Glasgow, Oxford, 26–28 September 1990
Geraint Jones and Mary Sheeran (Eds.)

Functional Programming, Glasgow 1990, Proceedings of the 1990 Glasgow Workshop on Functional Programming, Ullapool, Scotland, 13–15 August 1990
Simon L. Peyton Jones, Graham Hutton and Carsten Kehler Holst (Eds.)

4th Refinement Workshop, Proceedings of the 4th Refinement Workshop, organised by BCS-FACS, Cambridge, 9–11 January 1991
Joseph M. Morris and Roger C. Shaw (Eds.)

AI and Cognitive Science '90, University of Ulster at Jordanstown, 20–21 September 1990
Michael F. McTear and Norman Creaney (Eds.)

Software Re-use, Utrecht 1989, Proceedings of the Software Re-use Workshop, Utrecht, The Netherlands, 23–24 November 1989
Liesbeth Dusink and Patrick Hall (Eds.)

Z User Workshop, 1990, Proceedings of the Fifth Annual Z User Meeting, Oxford, 17–18 December 1990
J.E. Nicholls (Ed.)

IV Higher Order Workshop, Banff 1990
Proceedings of the IV Higher Order Workshop, Banff, Alberta, Canada, 10–14 September 1990
Graham Birtwistle (Ed.)